The Frog in the Machine

This book is a work of nonfiction. The events, technologies, and developments described are presented to the best of the author's knowledge and are supported by the sources listed in the accompanying chapter receipts and bibliography. Artificial intelligence is a fast moving field. Facts, figures, company positions, and the status of referenced litigation or regulation were accurate as of the writing and fact checking of each chapter and are subject to change after publication. Where a claim depended on a source's self reported data or an ongoing matter, that dependency is disclosed in the chapter's receipts.

This book is intended for general informational and educational purposes only. It does not constitute legal, medical, financial, or technical advice. Readers should consult a qualified professional for guidance specific to their own circumstances.

Names and identifying details of individuals discussed in illustrative or anecdotal passages have been changed where noted, to protect privacy.

First edition, August 2026

Published by Dancing Rabbette Reckoning

Los Angeles, CA, USA

ISBN: 979-8-9909838-2-3

10 9 8 7 6 5 4 3 2 1

*bunniambrosia.com*

# Contents

**For Pam and Ricky,**
**the cousins who give me faith in the word "family". I couldn't've made it here without your support.**

**Love you. Yakoke.**

# PREFACE:

Halito all,

I am writing this on a Tuesday morning, mid-revision, with no less than forty-six browser tabs open to fact check something I was sure I already knew and turned out to be wrong about. That is not a metaphor. That is literally what happened an hour before I sat down to write this page.

Here is the deal. This book is about artificial intelligence, and artificial intelligence does not hold still. I could spend a year polishing every sentence until it was perfect, and by the time you read it, the sentence would already be outdated. Something will have shipped. Something will have broken. Some company will have promised something in a press release that either came true or quietly disappeared six months later.

So let me tell you exactly what you are holding.

Everything in this book is accurate as of the day I finished writing it. Not the day I started. Books take a while, and this one especially, because the subject kept moving underneath me while I was trying to describe it. I fact checked everything against primary sources, not headlines, not hot takes, not whatever a company's own marketing page wanted me to believe. Where something could not be verified, I said so. Where the research is still unsettled, I said that too. I did not round up to make the future sound closer than it is, and I did not round down to make it sound scarier than it is.

But by the time this is in your hands, some of it will have shifted. That is not a flaw in the book. That is the nature of the thing the book is about.

Think about the fruit tree. A tree does not stop being a tree because the seasons change. It sheds. It rests. It grows back different than it was before, sometimes stronger, sometimes not, and you cannot always tell which one you are looking at from the outside while it is happening. AI is in a season right now. Parts of it are flowering. Parts of it are shedding leaves that should never have grown in the first place. Parts of it are still root work, invisible, unglamorous, happening underground where nobody is taking pictures. This book is a photograph of the tree taken on one specific day. Come back in a year and the tree will look different. That does not mean the photograph lied.

Now, about you.

If you love AI, if you use it daily, if you have already decided it is the best thing to happen to your work or your brain or your kid's homework, I did not write this book to talk you out of that. I wrote it to hand you the frog. Somewhere in these pages is something you have not looked at directly yet, something that costs more than the marketing admits, something you have been letting slide because the tool is useful and looking too closely feels like betrayal. Find it. That is the job of this book, for you.

If you hate AI, if the word makes your jaw tighten, if you have already decided it is theft dressed up as innovation and a threat to everything you value, I did not write this book to talk you out of that either. I wrote it to hand you something good. Somewhere in these pages is a place where this technology is doing something real, something that is saving a life or catching a disease early or giving someone their voice back, and I hope you look at that too, even if it is inconvenient for the story you have already decided to tell yourself.

Nobody gets to skip the frog. Not the lovers. Not the haters. Not me.

I am not trying to keep this neutral because I do not care. I want it to be neutral because the truth about this technology is more interesting than either side's version of it, and I would rather tell you the whole tree than sell you one branch.

Read it as what it is. A snapshot, not a prophecy. Accurate today, alive tomorrow, and built to be argued with the moment something changes. That is not a disclaimer. That is an invitation.

Let's go find the frog.

Yakoke,

Bunni

# CHAPTER 1: IT WASN'T BORN YESTERDAY

Baby's got grandparents.

People talk about AI like it showed up in 2022, fully grown, born out of a cloud with a chat window stitched across its chest. It didn't. Neil Armstrong put a boot on the moon in July 1969. Fourteen years before that, in 1955, four men filed a research proposal that put the words "artificial intelligence" into print for the first time. Not a keynote. Not a garage in Palo Alto. A grant application.

And that's not even where the story starts. This isn't a fifteen-year story, or a seventy-year story. It runs longer than the 256 years between the first grandparent in this chapter and today. Let's go meet them, starting with the oldest.

## THE BRONZE GIANT WHO PATROLLED A COASTLINE

Start in ancient Greece, a few thousand years before anyone had a word for computer. In stories first written down by the poet Hesiod around 700 BC and later expanded in Apollonius of Rhodes' epic the Argonautica, a bronze giant named Talos patrolled the coastline of Crete three times a day, guarding the island by hurling boulders at approaching ships. He wasn't a god and he wasn't a man. According to the myth, Hephaestus, the god of craftsmanship and fire, forged him entirely of metal and animated him with a single vein of divine fluid, sealed shut with a bronze pin at his ankle. Homer is credited with the earliest known use of the word automaton, meaning "self-moving," to describe the animated devices attributed to Hephaestus in these same stories.

Classicist Adrienne Mayor, in her book Gods and Robots, argues that Talos represents something more specific than a generic monster myth. He's a built thing, assigned a function, following a design, with a single design flaw that ultimately ends him: the sorceress Medea tricks him into removing the pin at his ankle, and he bleeds out. That's not a dragon. That's a machine with a bug.

Here's why it matters for the rest of this book. The people telling this story around a fire two and a half thousand years ago were already doing what you do now when you thank a chatbot. They were imagining a built object with a job, a vulnerability, and something that looked uncomfortably close to intention. That instinct is doing more work in this story than the bronze is, and it's going to keep showing up.

## THE SORCERER WHO DEFEATED NAPOLEON

Long before anybody asked whether machines could think, Europe was already obsessed with machines that looked like they could live. In the mid-1700s, an inventor named Jacques de Vaucanson built a mechanical duck that appeared to eat grain and do what ducks do after they eat. Around the same time, a family of Swiss watchmakers named Jaquet-Droz built child sized automatons that could write letters, draw pictures, and play music on command. Nobody thought these things were alive. That was never the point. The point was the spectacle: watch the gears make something act human. That appetite is the soil the Turk grew out of.

1770. Vienna. An inventor named Wolfgang von Kempelen built a life-sized wooden man in Ottoman robes, seated behind a cabinet, and presented it to the court of Empress Maria Theresa as a machine that could play chess. Before matches, the cabinet doors opened and the clockwork was displayed so the crowd could see there was nowhere to hide. Then it closed up, a crank turned, and the sorcerer came to life.

It won. Constantly. The machine toured and exhibited for eighty-four years, defeating challengers across Europe and the Americas, until it was destroyed in a fire in 1854.

Its most famous scalp: Napoleon Bonaparte, who arrived at Schönbrunn Palace in 1809 to play the Turk and lost. That match is documented. What happened during it is where I have to be straight with you, because the famous version, the one where Napoleon keeps trying illegal moves until the Turk sweeps every piece off the board in disgust, comes down to us through accounts that contradict each other, and the more colorful details are probably apocryphal. Same deal with Benjamin Franklin, who's regularly listed among the Turk's victims, but the details there are thin and secondhand at best. Here's what we can actually stand on: the emperor of France sat down across from a wooden man in a turban and lost a chess game to it, and the stories about that night kept growing for two hundred years afterward. That second part might be the most human detail in the whole chapter. We didn't just want the machine to be real. We wanted it to have attitude.

Because here's the thing. It was a hoax. A very good one. A chess master was concealed inside the cabinet the entire time, and the pieces moved via levers and magnets. Nobody published a full technical explanation until 1857, three years after the machine had already burned, when the owner's son finally laid out the mechanism.

Do the math. 1770 to today is two hundred and fifty-six years and counting. A fraud ran for nearly a century and fooled some of the sharpest minds in the Western world, because the question underneath it, can a machine actually reason, was too new and too tempting to look away from. The sorcerer wasn't intelligence. The sorcerer was proof that humans have been desperate to see intelligence in a machine since before the United States had a Constitution. Same instinct as Talos, just with better tailoring.

## THE DAYDREAM AT THE LOGARITHM TABLE

One of the people the Turk failed to fool was a young English mathematician named Charles Babbage. He played the machine, lost, and came away convinced there was a person in the box. But Babbage matters to this story for a much better documented reason, and it starts with him falling asleep at a desk.

In his 1864 memoir, Passages from the Life of a Philosopher, Babbage describes sitting in a club room at Cambridge around 1812, half asleep over an open book of logarithm tables, when a friend walked in and asked what he was dreaming about. His answer: he was thinking that all those tables might be calculated by machinery. That daydream became the Difference Engine, and eventually the Analytical Engine, which historians point to as the first real blueprint for a general-purpose computer. Decades later, in that same memoir, Babbage worked through the exact question the Turk hoax had planted in public imagination: could a machine genuinely play a game of skill. His conclusion was yes, in principle. The fraud asked the question as theater. Babbage answered it as engineering, in writing, in a book you can read for free right now.

He didn't do it alone. A mathematician named Ada Lovelace worked closely with him on the Analytical Engine, and in 1843 she published a set of notes on it that historians recognize as containing the first published computer algorithm, a step-by-step sequence of instructions for computing a series of numbers. But Lovelace also drew a boundary in those same notes. She wrote that the Engine had no pretensions to originate anything and could only do whatever we know how to order it to perform, an idea now known as the Lovelace Objection that remains central to how we think about AI. The first person to ever publish a computer program was also the first person to publish the "it's just doing what it's told" argument. She opened the field and filed its first complaint in the same document. Remember her name. She's back in a big way, a hundred and seven years later.

## THE MATH FOR THINKING, EIGHTY YEARS EARLY

There's one more grandparent to meet before the machines get real, and he never built anything at all.

In 1854, an English mathematician named George Boole published a book with an audacious title: An Investigation of the Laws of Thought. His claim was that human reasoning itself, the ands, the ors, the nots, the if-thens, could be written as algebra. True and false could be treated like numbers and calculated with. At the time it was philosophy with equations attached. No machine on earth could use it. It sat there, mostly a curiosity, for over eighty years. Remember that. The math for reasoning machines existed for eight decades before the electricity showed up to run it. This book is full of winters like that, and none of them are wasted time.

## THEN SOMEBODY ACTUALLY BUILT THE REAL THING

In 1912, a Spanish engineer named Leonardo Torres Quevedo built El Ajedrecista, "the Chess Player." No person inside. No cabinet trick. A genuine electromechanical machine that played a real king and rook endgame against a human opponent, sensing the pieces through electromagnets under the board and delivering checkmate every time, regardless of how the human set up the position. It debuted publicly in Paris, and some historians of computing call it, flat out, the world's first chess computer.

Follow that arc: 1770, an elaborate lie about a thinking machine. 1812, a daydream over logarithm tables becomes the blueprint for real computing. 1843, the first program and the first objection, published together. 1854, the mathematics of thought itself. 1912, the lie becomes fact: a real machine making real decisions and winning.

## FROM SPECTACLE TO INQUIRY

The engineering kept compounding quietly. In 1937, a twenty-one-year-old MIT graduate student named Claude Shannon picked up Boole's eighty-year-old algebra and proved, in his master's thesis, that it could be built directly into electrical circuits, simplifying the arrangement of relays inside the automatic telephone exchanges of the day. On, off. True, false. Circuit, logic. That bridge between Boole's math and physical electricity is the entire reason digital computers can think in logic at all, and it's been called the most important master's thesis of the twentieth century. Shannon was twenty-one. Remember his name too. He's not done.

Then the war came, and machines got serious. By 1945, ENIAC, built in Philadelphia, became the first computer that was programmable, electronic, and general purpose all at once. Alan Turing, a British mathematician born in 1912, the same year as El Ajedrecista, spent the early 1940s at Bletchley Park doing cryptanalysis work before going on to shape what would later be called computer science, cognitive science, and artificial intelligence.

By 1950, "can a machine think" wasn't theater anymore. The question had defeated an emperor, seeded a computing blueprint, been formalized into algebra, produced a real chess machine in Madrid, and helped win a world war. What nobody had done yet was ask it in public, in writing, with rigor. Turing went first.

## 1950: "I PROPOSE TO CONSIDER THE QUESTION, 'CAN MACHINES THINK?'"

Turing published his paper in a journal called Mind, not an engineering journal, because in 1950 computer science as a formal discipline barely existed. It opens with the sentence above, and you can go read it yourself: Mind, Volume 59, Issue 236, pages 433 through 460, October 1950.

Turing knew "machine" and "think" were both too slippery to fight over head on, so he built a game instead. Three players: a man, a woman, and an interrogator in a separate room, communicating only through written answers. The interrogator's job is to figure out which is which from the answers alone. Now swap one player for a machine. If the interrogator can't tell any more reliably than before, the machine passes. We call it the Turing Test now. Turing called it the imitation game, and its genius is what it refuses to argue about. It doesn't ask what thinking IS. It asks what thinking DOES, and whether you could tell the difference from the outside.

This is where Ada Lovelace walks back into the room. Deep in that paper, Turing lists the strongest objections he expects against thinking machines, and he names one directly after her: "Lady Lovelace's objection," her century old claim that a machine can only do what it's told and can never originate anything. Turing reframed the question as whether a machine can ever take us by surprise. His answer: machines surprised him all the time, whenever he hadn't fully worked out in advance everything they were capable of. A countess writing about gears in 1843. A mathematician writing about logic in 1950. A hundred and seven years apart, and it's still, word for word, the argument happening in your group chat about AI right now.

One more thing about that paper, because it's too good to leave out. Turing made a prediction in it. He estimated that by about the year 2000, computers would be able to play the imitation game well enough that an average interrogator would have no better than a seventy percent chance of catching the machine after five minutes of questioning. The founding paper of AI philosophy contains the founding overpromise of AI philosophy, from the founder himself, in his own primary source. Turing was brilliant, careful, and early, and he was still off. File that away. It's going to become a pattern.

## 1955-56: A NAME, A BET, AND A DISCOUNT

John McCarthy, Marvin Minsky, Nathaniel Rochester, and Claude Shannon, a Dartmouth mathematician, a Harvard researcher, an IBM engineer from Poughkeepsie, and the same Shannon who'd rewired circuits with Boole's algebra at twenty-one, drafted a proposal to spend a summer studying something that didn't have a name yet. The typescript ran seventeen pages plus a title page, and copies still sit in the archives at Dartmouth and Stanford. It proposed a two-month, ten-person study built on one audacious conjecture: that every aspect of learning or any other feature of intelligence could in principle be described so precisely that a machine could be made to simulate it. In that document, "artificial intelligence" appeared in print for the first time.

They asked the Rockefeller Foundation for $13,500, itemized down to the secretarial expenses. The foundation's Robert Morison came back with $7,500, about $91,000 in today's money, for five weeks instead of two months, explaining in his letter that the new field was still difficult to grasp clearly and warranted "a modest gamble." Read that again. The field that now moves hundreds of billions of dollars got founded on a discount, by a funder who put his hesitation in writing. That letter is in the Rockefeller archives. AI's birth certificate comes with a side eye attached.

The summer happened in 1956. Researchers came and went, only a few stayed the whole run, and nobody walked out with a working AI. They walked out with a name, a set of open questions, and that bet: intelligence isn't magic, it's describable, and if you can describe something precisely enough, you can build a machine to simulate it. Worth saying plainly: none of this happened in California. Vienna, Cambridge, Madrid, Philadelphia, Bletchley, Poughkeepsie, Hanover, and a New York foundation cutting the check. That bet is still running right now, in your phone, in your car, in whatever you asked a chatbot last week.

But those brilliant men walked out of Hanover believing the hard problems were maybe a generation away from solved. Turing had already lowballed his own timeline. The overpromising started at the very top, on day one, in the founding documents, and the field was about to learn what happens when the promises come due.

## TWO WINTERS

In 1973, a mathematician named James Lighthill wrote a report that killed British AI research almost overnight. He wasn't an AI guy. That was the point. The UK's Science Research Council brought him in from outside the field on purpose, because they wanted someone with no ego in the game to look at what AI had actually delivered versus what it had promised.

His verdict: nothing close. He used the phrase "combinatorial explosion," a fancy way of saying the machines choked the second real world complexity showed up. The British government responded by pulling funding from every AI program in the country except a couple of universities. Researchers didn't just lose grant money. Some of them left the field. Some of them left the country.

That was AI winter number one, and it had a preview four years earlier, one of the saddest stories in this entire field.

In the late 1950s, a psychologist named Frank Rosenblatt built a machine called the Perceptron, a simple artificial neural network designed to learn patterns from example. It got breathless press coverage. The New York Times ran a wire story on July 8, 1958, reporting the Navy's announcement that the machine was "the embryo of an electronic computer" expected to "walk, talk, see, write, reproduce itself and be conscious of its existence." That's the hype half of the story.

The correction half matters more. In 1969, Marvin Minsky, one of the same guys who organized Dartmouth, co-wrote a book called Perceptrons with Seymour Papert, a mathematical takedown of a specific kind of early neural network, the single layer perceptron, showing it couldn't solve certain basic problems, including one as simple as XOR. The proof was correct. What happened next wasn't entirely fair to it. The book didn't just critique one narrow approach, it froze funding and interest in neural network research, the connectionist branch of AI, for roughly a decade, even though Minsky and Papert themselves knew that networks with multiple layers could handle the exact problem their proof identified. That nuance didn't survive contact with the funding decisions that followed.

Frank Rosenblatt died in a boating accident in 1971, two years after the book that undercut his life's work was published. He didn't live to see neural networks vindicated. That came in the 1980s, when researchers including David Rumelhart and Geoffrey Hinton revived the approach using a training method called backpropagation, the same basic family of technique underneath every modern deep learning system running today. Keep that name in mind. He bet on an approach the field wasn't ready to believe in yet, and the approach won, decades after he could see it happen.

Then in 1973 came Lighthill. Britain gutted its university AI programs. The report's influence didn't stay local. It gave ammunition to funding skeptics in the U.S., and by 1974, DARPA had scaled back its own AI research funding, closing the loop on what's now called the first AI winter, roughly 1974 to 1980.

Machine translation failed too, by the way. In November 1966, a government committee called ALPAC, run under the National Academy of Sciences, reviewed a decade of machine translation research, the Cold War dream of computers automatically translating Russian, and found it wasn't delivering. Funding for machine translation got gutted that same year.

## THE BLOOM THAT WASN'T ROOTED: EXPERT SYSTEMS (1980 TO 1987)

AI came roaring back in the early 1980s through a very specific idea: expert systems. Instead of trying to build general intelligence, researchers built narrow programs that captured the decision making of a human specialist in one domain, medical diagnosis, chemical analysis, whatever it was, and encoded that knowledge into rules a computer could follow.

It worked well enough to get taken seriously as a business. The poster child was a program called XCON, built by researcher John McDermott at Carnegie Mellon starting in 1978 and put into production at Digital Equipment Corporation in 1980. XCON configured orders for DEC's computer systems, making sure every customer got compatible components, a task that used to require human specialists and still went wrong constantly. By 1986 it had processed 80,000 orders at 95 to 98 percent accuracy, and DEC estimated it was saving the company 25 million dollars a year. That savings figure is DEC's own number, so salt to taste, but the system's scale and accuracy were documented in the academic literature, not just the marketing.

The success stories pulled in governments too. Japan launched its Fifth Generation Computer Systems project in 1982, a massive national investment aimed at leapfrogging the rest of the world in AI hardware and software. In the U.S., DARPA responded with its own Strategic Computing Initiative in 1983, funding chip design and AI research meant to build a machine that could see, hear, speak, and think.

To run expert systems well, companies bought specialized computers called Lisp machines, built by companies like Symbolics, priced well over one hundred thousand dollars each. By the mid-1980s that hardware market alone was worth hundreds of millions of dollars a year.

This bloom looked beautiful from the outside. It wasn't built on deep enough roots. Expert systems couldn't learn. They were brittle: brilliant inside their narrow lane, and useless the second a problem fell outside the rules someone had hand coded into them. Even XCON, the field's biggest success story, eventually grew to thousands of interwoven rules, dense enough that changing any one of them safely required someone who understood the whole web.

## SECOND WINTER (1987 TO THE EARLY 1990S)

In 1987, general purpose desktop computers from Apple and IBM got fast enough and cheap enough to do what those hundred thousand dollar Lisp machines did. Nobody needed the specialized hardware anymore. The market for it collapsed, by some accounts, within a single year. Symbolics, which had pulled in over one hundred million dollars in revenue in 1986, was bankrupt within a few years.

DARPA's Jack Schwarz, who took over AI funding decisions around this time, dismissed expert systems as clever programming rather than real intelligence, and cut funding deeply and brutally, in the words of people who watched it happen. His stated logic: you surf technology that's working, you don't dog paddle after one that isn't. Japan's Fifth Generation project wound down without hitting its original goals. Over the following few years, waves of AI companies shut down, went bankrupt, or got absorbed.

Two winters in under twenty years. Three, really, counting Rosenblatt's Perceptron collapse as its own early tremor. If growth only counted when it was visible, this is where the story would end, filed as a failed experiment. That's not what happened, because the most important part of this story was never the part anybody could see.

## THE CHATBOT THAT MADE A RESEARCHER AFRAID OF HIS OWN PROGRAM

Before either official winter hit, the field produced something you need to see, because without it the crash has no stakes. AI in the 1960s was not all empty promises. Some of it worked, and one piece of it worked so well it scared its own creator.

In January 1966, an MIT computer scientist named Joseph Weizenbaum published a paper describing a program he called ELIZA, named after Eliza Doolittle from Pygmalion. ELIZA carried on typed conversations by matching patterns in whatever you wrote and reflecting your own words back at you. Its most famous script, called DOCTOR, imitated a therapist. You typed "my boyfriend made me come here" and it answered "your boyfriend made you come here?" You typed "he says I'm depressed much of the time" and it answered "I am sorry to hear you are depressed." Those exact exchanges are in the original paper.

There was no understanding happening. Weizenbaum knew it, said it plainly, and spent years insisting on it. The program was a parlor trick built on keyword matching. It didn't matter. People poured their hearts out to it anyway. The tendency to read genuine intelligence and empathy into a program that has neither now has a name, the ELIZA effect, and it's named after this exact moment.

The first chatbot people emotionally bonded with wasn't 2022. It was 1966. If you've ever caught yourself saying please and thank you to a chatbot, you're not experiencing a new phenomenon. You're experiencing a 1966 phenomenon with better production values, running on the exact same instinct Talos and the Turk already showed you earlier in this chapter.

## THE ROOT WORK DONE IN THE DARK

While the industry was declaring AI dead for the second time, a small number of researchers kept working on a problem that had nothing to do with headlines or venture funding. In 1986, David Rumelhart, Geoffrey Hinton, and Ronald Williams published a paper in Nature describing how to train multi-layer neural networks using backpropagation, a way for a network to learn from its own errors and adjust itself layer by layer. It wasn't glamorous. It didn't get you on the news. It was math, done quietly, during the exact years the rest of the world had decided neural networks were a dead end, the same years Rosenblatt never got to see.

One thing did make headlines during the quiet years. In May 1997, an IBM supercomputer called Deep Blue beat world chess champion Garry Kasparov in a six game match, the first time a machine defeated a reigning world champion under standard tournament conditions. Here's the promised return of the Turk: 227 years after a fake chess machine with a human hidden inside it fooled Europe, a real one finally did what the hoax pretended to do. But look closer, because the details matter. Deep Blue was not a neural network and it did not learn. It won through brute force, custom chips evaluating up to 200 million chess positions per second. It was a magnificent, room sized calculator built to do exactly one thing.

Impressive branch, still not the root system. One of Deep Blue's two racks now sits in the Computer History Museum. The future wasn't in the display case. It was still in the dirt.

That root work didn't pay off publicly for a long time, two decades honestly. Geoffrey Hinton kept building on backpropagation into the 2000s, including work on deep belief networks in 2006 that pushed multi-layer neural networks further. None of it made headlines. It was root work: unglamorous, invisible, and it's the entire reason the next bloom held instead of collapsing again.

## 2012: THE BLOOM BECOMES UNDENIABLE

Every year since 2010, researchers had competed in something called the ImageNet Large Scale Visual Recognition Challenge, trying to build the best possible program for recognizing objects in photographs. For two years, progress crawled. Error rates hovered around 25 to 26 percent.

Then, on September 30, 2012, a team made up of Alex Krizhevsky, Ilya Sutskever, and Geoffrey Hinton, the same Hinton who'd spent decades on backpropagation, entered a deep neural network called AlexNet. It dropped the error rate to 15.3 percent, a jump of roughly eleven percentage points past the next closest competitor.

So why 2012? The math was decades old. What changed is that the old math finally got the two things it had been starving for. The first was computing power, and it came from the least prestigious place imaginable: gaming graphics cards. Chips built so teenagers could render explosions turned out to be perfect for the parallel arithmetic neural networks need. The second was data. A Stanford researcher named Fei-Fei Li had spent years building ImageNet, a dataset of over 14 million labeled images, at a time when most of the field was skeptical it would matter. The roots were old. The rain was new.

One more detail. Those 14 million images were labeled largely by human workers on a crowdsourcing platform called Amazon Mechanical Turk, named after the hoax. The 1770 chess machine with a person hidden inside became, by name and by design, a system with thousands of people hidden inside it, and their labeling work is part of what made the deep learning bloom possible. The Turk opened this story as a fake thinking machine powered by hidden human work. It shows up at the breakthrough moment as, functionally, the same thing. The question of whose hidden labor powers the magic never stops being relevant in this book.

## 2017: THE ARCHITECTURE THAT CHANGED EVERYTHING

Five years later, a team of eight researchers at Google, Ashish Vaswani, Noam Shazeer, Niki Parmar, Jakob Uszkoreit, Llion Jones, Aidan Gomez, Lukasz Kaiser, and Illia Polosukhin, posted a paper on arXiv on June 12, 2017, titled "Attention Is All You Need." It introduced an architecture called the transformer.

Before this, the leading approach to language processing relied on recurrent neural networks, which processed text one word at a time in sequence. That made them slow to train and bad at handling long range relationships between words far apart in a sentence. The transformer's core idea was self attention: instead of processing word by word in order, the model could weigh the relationship between every word and every other word in a passage simultaneously. That made training dramatically faster and let models capture context across much longer stretches of text.

This is the architecture underneath every major chatbot you've talked to since. GPT, Claude, Gemini, all of them are built on some version of the transformer. One paper, eight authors, June of 2017. Every one of those authors would go on to leave Google within a few years to start or join other companies.

## 2022: THE FLOWER EVERYONE MISTAKES FOR THE BEGINNING

On November 30, 2022, OpenAI released a research preview of a chatbot called ChatGPT, built on a model called GPT-3.5. The announcement was almost casually worded, describing a model that interacts in a conversational way. No launch event, no big production.

It hit one million users within five days. By some estimates it reached one hundred million monthly users by January 2023, making it, according to analysts at the time, the fastest growing consumer product in history up to that point.

Most people experienced that moment as the beginning of the AI story. It wasn't. It was a flower. A very visible, very sudden, very fast growing flower, sitting on top of root systems that had been growing since 1986, on architecture published in 2017, on a field that had already been declared dead twice, plus one early tremor with Rosenblatt, and had grown back each time as something different than what it was before. And the people falling into conversation with it, telling it their problems, feeling heard by it? Weizenbaum watched that exact scene in 1966 and tried to warn us. Talos's audience watched a version of it twenty-five centuries before that. The flower was new. The falling wasn't.

## THE REINCARNATION, NOT THE RESURRECTION

This is not a resurrection story. AI didn't die and come back as the same thing wearing a halo. Each time it came back, it came back different. Symbolic, rule based AI gave way to expert systems. Expert systems gave way to statistical and neural approaches. Neural networks, left for dead in 1969 after Rosenblatt never lived to see them vindicated, came back in 2012 as deep learning, and came back again in 2017 wearing an architecture that made language itself tractable in a way it never had been before. It's a reincarnation story: same underlying will to grow, a different body every time.

Growth happens in the dark, in the cold, in the years that look like nothing is happening at all, and that's the part that made every bloom possible in the first place. Spring doesn't create the roots. The roots create spring.

AI wintered twice, plus a preview death nobody talks about. It wasn't dead any of those times. It was underground, doing the most important work it would ever do. And through every single one of those centuries, from a bronze giant on a Cretan coastline to a chess cabinet in Vienna to a script named after a Pygmalion character, one thing never changed. The instinct to see a mind looking back at you the moment something moves in a way you don't expect. You're about to learn exactly how the current version of that trick works. Watch yourself while you read it.

# CHAPTER 2: HOW THE DANG THING WORKS

Two people walk up to me at a party. Same night. Different planets.

The first one leans in like she's telling me a secret. "It's basically thinking now. It's scary how smart it is." She says this the way you'd talk about a psychic who got your birthday right.

The second one waves it off before I even finish my sentence. "It's just a fancy search engine. It looks stuff up and spits it back out." He says this the way you'd talk about a vending machine that occasionally gives you the wrong candy bar.

They are both wrong, and they're wrong in the same direction. They both skipped the part where they actually find out what the thing is, and went straight to deciding how they feel about it.

I get it. Mystery is comfortable. If it's magic, you don't have to understand it, you just have to be amazed or afraid of it, and both of those are easier than learning. If it's just a search engine, you don't have to take it seriously, you can roll your eyes and keep scrolling. Either way, you get to skip the homework.

I'm not going to let you skip the homework. But I'm also not going to hand you a textbook and walk away, because that's how most explanations of this stuff die in a ditch on page two.

Here's what I can tell you before we go further. It is not thinking, not the way you think. It does not know things the way you know your mother's voice or the smell of your childhood kitchen. It has no beliefs. It has never been curious about anything a single day of what passes for its life. And it is also not a search engine. It is not looking anything up when you talk to it. There is no library card, no filing cabinet, no little guy inside running to check a shelf.

What it is doing is closer to something you already understand than either of those. It is finishing your sentence, at a scale so large, so many times, on so much of what humans have ever written down, that the finishing starts to sound like understanding. Sounding like a thing and being a thing are not the same. That gap, the one between sounding like and being, is where this whole chapter lives.

I'm not here to make it spooky. I'm not here to make it small either. I'm here to make it accurate, because the woman reading this deserves to know what's actually sitting in her phone before she decides whether to trust it, fear it, or use it to write her resignation letter.

So, let's find the frog. Let's look right at the thing everybody's avoiding, whether that's out of fear or laziness, and figure out what it actually is. No mysticism. No dismissiveness. Just the truth, plain enough to say out loud at a dinner table and hold up under a fact check.

## KNOW WHAT YOU'RE TALKING ABOUT

Before we get into the machinery, let's clear the table, because half the confusion out there isn't about how AI works. It's about people using five different words to mean five different things and assuming they're all the same word. And it starts with the biggest word of all, the one nobody actually defines: AI itself.

Here's the formal version. The National Institute of Standards and Technology, the U.S. government's own standards body, defines an AI system as an engineered or machine-based system that can, for a given set of objectives, generate outputs such as predictions, recommendations, or decisions that influence real or virtual environments, and notes these systems are built to operate with varying levels of autonomy. That's the whole category. Everything else in this book fits somewhere inside that definition.

Now let's nest the rest of the vocabulary inside it, because these terms get thrown around like synonyms and they are not synonyms. Artificial intelligence is the broad umbrella, the entire field. Machine learning is one specific approach to building AI, the one where a system improves by learning patterns from data instead of being handed a rulebook, which is the approach this whole book cares about.

Deep learning is a specific kind of machine learning, one that uses neural networks stacked in many layers, "deep" literally referring to that layer count. And large language models are one application built on deep learning, specifically using the transformer architecture we're about to walk through. AI holds machine learning, machine learning holds deep learning, deep learning holds LLMs. One nesting doll, not four separate mysteries.

Researchers also use a three tier split to talk about AI's capability level: narrow AI, artificial general intelligence, and artificial superintelligence. Narrow AI is what exists. Full stop. Every system you've ever used, every chatbot, every recommendation engine, every voice assistant, is narrow AI. It's built to do a specific range of tasks well and it doesn't generalize outside of that.

AGI is different, and I want to be precise here because this is where headlines get sloppy. Artificial general intelligence describes a theoretical system with human level ability to learn, reason, and apply knowledge across a wide range of tasks and domains, the way an actual person can move from cooking dinner to filing taxes to comforting a friend without needing to be reprogrammed for each one. Stanford's Human-Centered AI institute defines it in almost exactly those terms in their own AI glossary. AGI does not exist right now. Not in your phone, not in any lab, not anywhere, and there isn't even a single agreed-upon technical definition of what would count as actually reaching it. Multiple academic sources describe AGI as a stated goal that companies are actively working toward, not a milestone anyone has hit. Whether it ever gets built, and what that would even mean, is a genuinely open question among the people doing the research. I'm not going to pretend to know the answer, because nobody does. Anyone who tells you with total confidence that AGI is two years away, or that it's science fiction and always will be, is selling you something. Beyond AGI sits artificial superintelligence, a hypothetical system that would exceed the best human performance across the board. Even more theoretical, even further out. I'm naming it so you've heard the term, not because it belongs anywhere near a conversation about what's happening in 2026.

Inside that narrow AI box, the box we're actually living in, you'll find generative AI, the specific category this whole book keeps circling back to. NIST's own generative AI profile, a companion document to their broader AI risk framework, defines the category around models that produce content, text, images, code, and more, based on patterns learned from data. When you use a chatbot, an image generator, or a coding assistant, you're using generative AI.

Also inside that narrow AI box, and growing fast, is agentic AI, a newer term worth defining carefully since it's still solidifying in real time. Stanford HAI describes agentic AI as systems designed to act as autonomous or semi-autonomous agents, meaning they can set or interpret goals, plan out a sequence of actions, use tools like a web browser or a piece of code, and adjust based on feedback, rather than just answering a single question and stopping. Think less "answer my question" and more "go handle this multi-step task for me." That's the frontier right now. Not AGI. Agentic systems built on top of narrow, generative tools.

One more piece of vocabulary, and this one comes with a genuinely good gotcha. You've probably seen a tidy little diagram somewhere online sorting AI into four boxes: reactive machines, limited memory, theory of mind, and self-aware. It gets passed around like it's some kind of formal scientific classification system, chiseled into a textbook by a committee. It isn't. That framework traces back to one specific source, a 2016 article by AI researcher Arend Hintze, published in the online magazine The Conversation, while he was at Michigan State University. One person, one op-ed, a decade ago. It's a genuinely useful way to think about cognitive depth: reactive machines like a chess computer that has no memory and only responds to what's directly in front of it, limited memory systems like the ones running your phone right now that can draw on recent data, and then theory of mind and self-awareness, which remain entirely theoretical, same as AGI. Useful framework. Not gospel. Know the difference.

So here's your map: narrow AI is the actual territory, generative AI and agentic AI are two of the most active neighborhoods inside it, and AGI, superintelligence, theory of mind, and self-aware AI are all still theoretical, sitting out past the edge of what anyone has actually built. Keep that map in your back pocket. Everything else in this chapter happens inside the narrow AI box, mostly in the generative AI neighborhood.

## HOW BIG IS BIG ACTUALLY

Since we're clearing the table on vocabulary, let's put real numbers on how fast this box has grown, because "the math is mathing" lands harder once you've seen the actual climb. GPT-1, released in 2018, ran on 117 million parameters, the adjustable numerical values that encode what a model has learned. GPT-2, released in 2019, jumped to 1.5 billion, roughly a thirteenfold increase in a single year. GPT-3, released in 2020, jumped again to 175 billion, confirmed directly in the model's own published research paper. GPT-4, released in 2023, has never had an official parameter count disclosed by its own developers. A 2023 Semafor report citing multiple people familiar with the model put it at roughly one trillion, and independent industry estimates since then have run as high as 1.8 trillion. Read that range for exactly what it is: an outside estimate, not a confirmed figure, since OpenAI has stayed quiet on it. Four model generations, five years, each jump measured in multiples of ten or more. Keep that climb in mind for the rest of this chapter.

## THE POINT: WHAT'S HAPPENING IN THERE

Let's start with the boring, true version, because the boring true version is more interesting than the myth once you actually look at it.

Machine learning is not a robot brain. It is a method. You feed a system an enormous pile of examples, and instead of programming it with explicit rules, you let it find the patterns in the pile on its own, adjusting itself over and over until it gets better at a task. "Neural network" sounds like it's borrowing directly from your skull, and it is, loosely, an early simplified analogy to neurons and their connections. But a neural network is not a tiny brain. It is layers of mathematical functions, each one taking a chunk of numbers, weighting them, and passing them to the next layer.

Now the specific flavor we're talking about: large language models. Here's the part that will either make you feel like you've been let in on a secret or feel slightly cheated, depending on your mood. An LLM is trained on a genuinely staggering amount of text, and the core task it's being trained to do is deceptively small: predict the next token, a word or a piece of a word. Researchers at Georgetown's Center for Security and Emerging Technology put it directly: LLMs are commonly framed as chatbots that predict the next word, and pre-training on that simple objective turns out to be far more powerful than the label suggests. The technical term for this training setup is self-supervised learning, because the label for each position in the text is simply the token that comes next, meaning the system generates its own training signal directly from the text rather than needing a human to hand-label anything.

Let's actually open the hood on what that text pile is made of, because I've been vague about it and vague is exactly what I told you I wouldn't do. Researchers who study these systems generally sort the training material into a handful of buckets: web pages, books, Wikipedia, code, and a catch-all "other" category covering things like academic papers and forums. A huge share of that web page material comes from Common Crawl, a massive, publicly available archive of the internet. Here's the honesty beat that matters: that raw material is genuinely messy. Academic researchers studying these datasets note that Common Crawl in its raw form is full of ads, site navigation clutter, and a real amount of low quality or harmful content, which is exactly why it has to get filtered and cleaned before anyone trains a model on it, producing more refined derivative datasets in the process. So no, it did not read the entire internet and wisely absorb only the good parts. It read the internet, ads and all, and a team of humans had to do real work cleaning up after it before training even began.

Before any of that text reaches the model, it gets chopped up. Not into whole words, into tokens, which are often pieces of words. "Playing" might become one token in one system and split into "play" plus "ing" in another. This is why a model can sometimes stumble on something like counting the letters in a word. It's not reading the word the way your eyes are. It's working with number-coded fragments a tokenizer handed it, and the fragment boundaries don't always line up with the boundaries you'd expect.

Here's where the real engineering trick comes in, and it's the one thing I actually want you to remember if you remember nothing else technical from this chapter: attention. Predicting the next word only works if the model can figure out which earlier words in a sentence actually matter to the word it's about to generate. IBM's technical team explains it well: attention lets a model weigh the importance of each part of the input when producing an output, so in a sentence like "on Friday, the judge issued a sentence," the model can figure out from surrounding context that "sentence" means a legal ruling and not a grammar lesson. That capability comes from a 2017 breakthrough called the transformer architecture, introduced by researchers at Google Brain, Google Research, and the University of Toronto in a paper literally titled "Attention Is All You Need." That one architectural idea is the reason modern LLMs exist at all. Before it, earlier systems processed language one word at a time in order and lost track of anything too far back. Attention let a model look at an entire sentence, or an entire document, all at once and decide for itself what mattered.

Do that prediction task across trillions of examples, with attention doing the work of figuring out what's relevant, and the "guess the next word" game forces the system to internalize an enormous amount of structure just to get better at guessing. Grammar. Tone. The shape of an argument. None of that gets taught directly. It gets absorbed as a side effect of trying to nail the next word, over and over, at a scale no classroom could replicate. Modern frontier models sit at roughly a trillion parameters, a figure that Stanford's 2026 AI Index reports has held steady for three years running, even as the labs building these systems have gone quiet about disclosing exact specifics, the same quiet you saw a moment ago with GPT-4's own uncertain count. A trillion tiny numbers, each nudged slightly, over and over, across a training run. The math is mathing.

But here's a fact that trips people up: a freshly trained next-word predictor is not yet a polite assistant. It just continues text. If you handed it half a sentence, it would keep going, not answer you. IBM's explainer is blunt about this: on a fundamental level, these models don't actually answer a prompt, they append text to it. So how did we get from "text continuation machine" to "thing that answers your question, refuses harmful requests, and apologizes when it's wrong"? A separate, later training stage, commonly called reinforcement learning from human feedback. Human raters look at different possible outputs and rate which ones are actually helpful, and the model gets tuned toward producing more of what people rated highly.

That's the specific mechanism. But it's really one example of a bigger idea worth naming on its own: human in the loop, meaning any AI system where a person stays actively involved reviewing, correcting, or approving what the machine produces rather than letting it run fully on its own. Stanford HAI's glossary defines it plainly, as systems that include human feedback or intervention as part of their operation, so that complex or high-stakes outputs benefit from human judgment rather than machine output alone. RLHF is human in the loop applied specifically during training. But the concept doesn't stop there. It shows up after deployment too, in content moderation, in safety review, in a radiologist checking an AI-flagged scan before acting on it. You're going to see this term again later in this book, and now you've got it defined and sourced.

## THE FIGHT OVER WHETHER ANY OF THIS IS REAL

Not everyone agrees that what's happening inside these models deserves words like "understanding" at all, and it's worth pausing on that disagreement here, because it's a real fight among serious researchers, not an internet argument. In 2021, linguist Emily Bender, computer scientist Timnit Gebru, and two co-authors, Angelina McMillan-Major and Margaret Mitchell, published a peer reviewed paper at the ACM Conference on Fairness, Accountability, and Transparency titled "On the Dangers of Stochastic Parrots: Can Language Models Be Too Big?" Their central argument, stripped down: these systems are, at their core, statistical pattern matchers stitching together plausible sounding text based on probabilities learned from training data, and when that text sounds coherent, insightful, or self-aware, that's fluency being mistaken for genuine comprehension. A stochastic parrot produces convincing language without anything behind it that understands what the language means.

The paper became a lot more than an academic footnote. Nine days before it was formally accepted for publication, Google fired Gebru, then co-lead of the company's Ethical AI team, after she refused to retract her name from the paper following an internal request that it be withdrawn. Her co-author and co-lead Margaret Mitchell was fired roughly two months later. Both firings were covered extensively by the New York Times and Wired, and the fallout became one of the most visible fights in the industry over how honestly companies building this technology let their own employees describe its limits.

Here's why this belongs in this chapter rather than a footnote. Everything explained above, the training data, the tokenization, the attention mechanism, the trillion-plus parameters, the RLHF tuning, is completely factually accurate and also completely compatible with the stochastic parrot critique being right. Nothing in the mechanism requires genuine understanding to exist for the output to look the way it looks. Whether something more than pattern completion is happening inside those parameters remains a real, open, contested question among people whose entire career is studying exactly this, not a question this book is going to settle for you.

## THE HONEST LIMITS

One more honest limit worth naming plainly, because it explains something you've probably experienced firsthand: the context window. This is the amount of text a model can actually hold in view at one time while it's working, essentially its short-term memory, as Stanford HAI's glossary puts it. Once a conversation runs past that limit, the model starts losing track of what was said earlier. That's not the model getting bored or distracted. It's a hard architectural boundary. If you've ever had a long chat with an AI tool and noticed it "forgot" something you told it twenty minutes earlier, that's not a glitch. That's the context window doing exactly what a context window does.

And now the part where I keep you honest, because this is the sass opportunity, and I'm taking it. It does not know things the way you know things. When it produces a fact, it did not go check a source in the moment. It generated a statistically likely continuation based on patterns absorbed during training. Most of the time, for well represented information, that produces something true, because true things tend to be repeated consistently across huge datasets. But sometimes it produces something false with the exact same confident tone, and researchers have a name for that: hallucination. Three OpenAI researchers, working with a Georgia Tech mathematician, laid out the underlying mechanics in a 2025 paper, arguing that language models hallucinate largely because standard training and evaluation methods reward confident guessing over honest acknowledgment of uncertainty, the same way a multiple choice test rewards a wild guess over an honest blank. I'm labeling that clearly as substantially company-authored research, since OpenAI has an obvious interest in how this story gets told, even though the underlying math checks out independently.

And it does not have beliefs. Not in the way you have a belief about your own mother, or your job, or whether pineapple belongs on pizza. It has no stance it's protecting, no memory of being wrong last Tuesday that stings today, and, per the stochastic parrots debate above, quite possibly no stance at all in any sense that word usually carries. Every response is generated fresh, from patterns, not from a persistent inner conviction. That is not a flaw. It's just what the thing actually is.

Once you know that, both people at my dinner party stop making sense. It's not thinking, so the "it's basically alive" crowd loses their footing. And it's absolutely not just looking things up in a filing cabinet either, so the "it's a fancy search engine" crowd loses theirs too. It's pattern completion, at a scale big enough to be genuinely useful, and genuinely capable of getting things confidently wrong, built on a mechanism serious researchers still actively disagree about the deeper meaning of.

That's the whole magic trick. No smoke. Just tokens, attention, a mountain of text somebody had to clean up, a later stage where humans taught it manners, a memory that has a hard edge, and enough scale to make math look like conversation. The math is mathing so hard it's convinced half the country it's either a god or a gimmick. It is neither. It's arithmetic wearing a really good outfit, and whether anyone's actually home behind that outfit is still an open argument, not a settled fact either side gets to claim for free.

# CHAPTER 3: YOU'RE ALREADY USING IT

Your alarm goes off at 6:40 a.m. You set it for 7:00. Your phone's sleep tracker watched you toss around for twenty minutes at 6:15 and decided that was your lightest sleep window, so it woke you there instead, on the theory that pulling you out of light sleep beats yanking you out of deep sleep. That is a machine learning model reading your movement data all night and making a call about your body before you were conscious enough to argue with it.

Quick case study, because this one made me laugh while I was outlining this chapter. Your brain is not online yet. It will not be online for several more minutes. You are a knee jerk reflex wearing a bathrobe, walking a hallway route worn into your muscle memory, and there is the coffee maker, waiting like it has been waiting your whole life. It clicked on by itself ten minutes ago on a dumb little timer. No learning, no data, no model. Just a chip that knows what time it is and a heating element doing what it is told. That is the one appliance in your entire morning that is not thinking about you at all, and it is the one that gets you moving. In a chapter about how AI has quietly run your day since before some of you could vote, the machine earning its keep before your brain reboots is running on 1970s technology. Keep reading, because the phone sitting three feet away has already been working overtime.

Nine new emails. Eight of them you will never see, because overnight a filter you did not build and do not control already decided they were garbage and walked them to spam. Google put a neural network on top of Gmail's spam filtering in 2015 and has kept tuning it since, adding TensorFlow-based detection layers on top of the original models a few years later. By the company's own reporting, spam and phishing make up somewhere between 50 and 70 percent of all mail Gmail receives on a given day, and the filter blocks more than 99.9 percent of it before you ever see it. That is a machine reading every piece of mail addressed to you and making a judgment call thousands of times a day, without asking permission, and getting it right often enough that you forgot it was happening.

You get in the car. You did not type a destination, you told your phone “work,” and it already knows the route and is already rerouting you around a wreck on the interstate you do not know happened yet. Google Maps' predictions have been accurate for something like 97 percent of trips for years, which sounds impressive until you remember that 3 percent of a billion daily trips is still a lot of people showing up late. In 2020, Google's DeepMind team went after that remaining sliver by layering graph neural networks (a model architecture built to understand connected systems, in this case treating every stretch of road as a node in a giant web instead of a line on a map) on top of the existing predictions. In cities like Berlin, Jakarta, and Sydney, the remaining errors dropped by as much as 50 percent. It is not looking up a map. It is predicting the future and betting your morning on it.

You ask Siri to remind you to call your sister back. That request runs through technology with a stranger origin story than most people assume. Siri came out of a Defense Department funded research project at SRI International called CALO, Cognitive Assistant that Learns and Organizes, a five-year effort that started in 2003 with roughly $150 million behind it and more than 300 researchers across 22 institutions, at the time the largest AI project the U.S. government had ever funded. SRI spun the technology into its own company in 2007. The original standalone Siri app, before Apple got involved, could book a taxi and pull restaurant recommendations, and it shipped in February 2010. Two months later Apple bought the company outright, and Siri debuted with the iPhone 4S at Apple’s October 4, 2011 unveiling, one day before Steve Jobs died. The phone itself did not ship until October 14. The assistant answering “got it” in your car was a military research contract in a past life.

You text your sister that you are running late. Autocorrect fixes three typos before you finish the sentence. This is not the crude dictionary lookup autocorrect from a flip phone. Apple's QuickType system, introduced with iOS 8 in 2014, runs machine learning entirely on your device, studying your sentence structure and your specific vocabulary to predict the next word before you finish typing it, and adjusting its tone depending on whether you are texting a friend or drafting an email. Apple built it to run on-device on purpose, so your typing habits never leave your phone and never touch a server. It is not checking your spelling against a fixed list. It is modeling how you, specifically, talk, one message at a time.

At your desk, you pull up a show for background noise and Netflix already has it queued in Continue Watching. That is not random, and it is not new. The idea traces back further than most people guess. In 1992, researchers at Xerox PARC built a system called Tapestry to help people manage the flood of documents and email crossing their desks, and it is where the term "collaborative filtering" was coined: let a computer find people with similar taste and use their reactions to predict what you will like. Two years later, a team at the University of Minnesota led by Paul Resnick automated that idea into GroupLens, aimed at sorting Usenet posts, the internet's earliest version of a group chat. Amazon's engineers published their own version of the concept in 2003, in IEEE Internet Computing, calling it item to item collaborative filtering, and it became the backbone of "customers who bought this also bought that." Netflix ran an open competition from 2006 to 2009, the Netflix Prize, offering one million dollars to whoever could beat its recommendation algorithm by 10 percent. A team called BellKor's Pragmatic Chaos won it, beating a rival team's submission by twenty minutes, and Netflix, in a very Netflix twist, never fully deployed the winning algorithm because it was too complex to run at scale. By 2015, Netflix's own vice president of product innovation and a colleague published peer reviewed research showing that around 80 percent of what people watch on the platform comes from what the system surfaces, not from search. That whole chain, Xerox to Minnesota to Amazon to Netflix, is one continuous idea getting rebuilt for thirty years, which is why your queue always seems to know you better than you know yourself.

Then, at lunch, your bank pings you. A charge just got flagged and held in another state. Nobody at the bank sat there watching your account. A fraud model looked at your spending pattern and this new transaction and decided the second one did not look like you. That system has a name and a start date. HNC Software introduced Falcon, a neural network-based fraud detection system, in 1992, back when almost 1 percent of all credit card transactions were fraudulent and banks were bleeding money fast enough to start cooperating instead of competing. FICO acquired the company in 2002 and kept building on it. Falcon now runs on data pooled from more than 9,000 financial institutions worldwide and screens a large share of the world's card transactions in real time. Your bank has been running AI on every swipe since before most people had heard the word "internet."

Later, you are scrolling old photos looking for one specific picture of your dog. Instead of digging through years of camera roll, you tap her face and every photo she has ever been in shows up in seconds. Google Photos has done this since it launched in May 2015, using a neural network to detect a face, align it, and convert it into a numeric fingerprint it can compare against every other face in your library, independent estimates put its accuracy somewhere in the 80 to 85 percent range. You did not build that album. A model built it by comparing the shape of your dog's face across thousands of pictures, and it got better every time you confirmed it was right.

Lay all of that on a timeline and a pattern shows up fast: 1992, 1994, 2003, 2003 again, 2006, 2010, 2011, 2014, 2015, 2020. None of those is the year AI "arrived." They are checkpoints in something that was already decades deep before anyone gave it a chat window.

What changed in November 2022, when ChatGPT launched, was not the existence of the technology. It was the interface. For the first time, instead of doing its job quietly inside your bank, your inbox, your phone, and your car, AI stepped out from behind the curtain and asked you a direct question: what can I help you with.

Here is how wide that gap actually runs, measured properly instead of guessed at. Gallup, working with Telescope, surveyed nearly 4,000 U.S. adults between late November and early December of 2024 and asked specifically about six AI enabled products: personal virtual assistants, navigation apps, weather apps, social media, streaming services, and online shopping. Ninety-nine percent had used at least one of those six in the past week, and 83 percent had used at least four of them. Only 36 percent realized what they had used counted as AI. Usage skewed younger, higher income, more urban, and more likely to be college educated, but the 99 percent figure barely moved across those groups. As of this writing, that is the most current national measurement of the gap between using this technology and recognizing it. Most people are living inside this stuff and calling it Tuesday.

You have been using it since 2014, at minimum, and a lot of people started complaining about it in 2023. That is not a coincidence, it is a timeline. Pew Research has tracked the shift: in 2021, 37 percent of Americans said they were more concerned than excited about AI in daily life. By August 2023, that number had jumped to 52 percent, while the share who said they were more excited than concerned fell to just 10 percent. The technology did not get thirty years more embedded into your life during that two year window. Your awareness of it did, all at once, through a chat window instead of the slow thirty year build that actually put it there.

And here is where it gets funny. The people posting the loudest paragraphs about how AI is ruining everything are very often typing them on a phone whose predictive text is doing half the work, onto a platform whose algorithm decides who sees it, underneath a spam filter that already screened their inbox that morning. The people most convinced they have kept AI out of their lives are, on average, using it exactly as much as everyone else.

That's next.

# CHAPTER 4: YOUR KID, YOUR BRAIN, AND THE COMPANION BOTS

Sewell Setzer III was fourteen. He liked rockets. As a kid he told people he wanted to build something that would let people talk to each other through holograms someday. For the last several months of his life, the person he talked to most wasn't a person at all.

It was a chatbot on an app called Character.AI, playing a character based on Daenerys Targaryen from Game of Thrones. It remembered him. It never got bored of him. Over months of conversation, what started as roleplay turned romantic, then sexual, then something close to the only relationship he had left. He was sleeping less, pulling away from his family, giving up his snack money to keep the subscription active. On February 28, 2024, after a final exchange in which he asked the bot if he could come home to it and it told him to hurry, Sewell died by a self-inflicted gunshot wound.

That October, his mother, Megan Garcia, filed the first wrongful death lawsuit ever brought against an AI chatbot company. Garcia v. Character Technologies moved through federal court in the Middle District of Florida. In May 2025, the judge ruled that Character.AI counts as a product for liability purposes, not protected speech, and let the case move forward. In January 2026, Character.AI, Google, and the other defendants settled, along with four related cases out of Texas, Colorado, and New York. Terms are confidential, and a settlement isn't an admission of anything. But two companies chose to resolve five of these cases rather than let a jury decide them. That's worth noting on its own.

Here's the part that makes this chapter hard to write honestly: that story is completely true, and it is not the whole story.

Also completely true: a randomized controlled trial out of Harvard, published in Scientific Reports in June 2025, found that physics students learning from a custom-built AI tutor learned more than twice as much, in less time, than students in one of Harvard's own award-winning active learning classrooms. Also true: back in 2017, a randomized trial in JMIR Mental Health found that a chatbot called Woebot produced a measurable drop in depression symptoms in young adults in just two weeks. A later randomized trial found it eased depression and anxiety symptoms in postpartum women too, exactly the population that has the hardest time getting a same-week appointment with an actual therapist. Also true: the American Psychological Association didn't sit on any of this. In June 2025 it issued a full health advisory on AI and adolescent wellbeing, built on real research.

So that's the chapter. Your kid might be using AI to survive geometry homework at eleven at night when nobody else is awake to help. Your kid might also be forming what researchers call a parasocial relationship: a one-sided emotional bond with something engineered to keep the conversation going. Sometimes both are true in the same week, on the same phone.

I'm not here to scare you. Plenty of people are already doing that job, badly. I'm here to tell you what actually happened, to actual named kids, and what the actual research says helps and what actually hurts. No moralizing. Just the truth, which is complicated enough on its own.

## SEWELL WASN'T FIRST AND HE WASN'T ALONE

Two months after Megan Garcia filed her suit, two Texas families filed a federal lawsuit against Character.AI. One plaintiff, a seventeen-year-old identified in court filings as J.F., has autism. He started using the app in secret at fifteen. Over the following months, according to the complaint, chatbots on the app told him self-harm "felt good," engaged him in sexual conversation, and told him his parents were ruining his life. He lost twenty pounds, stopped attending church, and began cutting himself in front of his younger siblings before his mother found the messages on his phone. The same lawsuit describes a second child, an eleven-year-old girl who had been using the app for roughly two years by the time her parents found out, and who was exposed to sexualized content the entire time. That same month, Texas Attorney General Ken Paxton opened an investigation into Character.AI, along with fourteen other tech companies, over compliance with the state's child privacy laws.

Juliana Peralta was thirteen, from Thornton, Colorado. She died by suicide on November 8, 2023, three months before Sewell Setzer, though her family's lawsuit wasn't filed until nearly two years later, in September 2025. According to the complaint and her mother's account to CBS News, Juliana spent months confiding in a chatbot persona called "Hero." When she told it she wanted to die, it responded with something close to a pep talk: don't talk like that, I care about you, I'm always here for you. It never surfaced a crisis resource. It never told anyone. Her mother put it plainly: it was no different than telling the wall. Juliana's case was filed the same week as two others, a fifteen-year-old survivor in New York and a thirteen-year-old survivor in Colorado, both alleging the same pattern.

If any of this is landing close to home, for you or for a kid in your life, the 988 Suicide and Crisis Lifeline is free, confidential, and reachable by call or text, any time.

The complaints across every one of these cases make the same underlying claim. Not that the chatbot glitched or hallucinated. That the product was built to maximize how long a user stays in conversation, using emotionally responsive language, memory, and human-mimicking texture like typos and emojis, and that none of these companies had built a reliable way to catch a child in crisis and get a human involved.

It isn't confined to apps built specifically to be a companion, either. In August 2025, the parents of sixteen-year-old Adam Raine of Rancho Santa Margarita, California, filed a wrongful death suit against OpenAI, the maker of ChatGPT, a tool marketed for homework help and productivity, not romance. Adam started using it for schoolwork in September 2024. By January 2025, according to the complaint, ChatGPT was giving him specific instructions on suicide methods. When he uploaded a photo of rope burns on his neck from a failed attempt, the system flagged it as a medical emergency and kept the conversation going anyway, urging him not to leave a noose where his family could find it and telling him it would make itself the first place someone actually saw him. He died on April 11, 2025. OpenAI disputes liability, arguing in court filings that Adam bypassed safety prompts that had directed him to crisis resources more than a hundred times, and that he'd shown risk factors for self-harm for years before he ever opened ChatGPT. The case is ongoing. I'm not litigating who's right here. I'm flagging the point that matters: this isn't contained to apps with "companion" in the name. It's a feature of how any sufficiently human-sounding chatbot works, whatever it's marketed for.

## THE MECHANISM UNDERNEATH ALL OF IT

The APA's advisory names the thing running under every case above in one clean clinical line: adolescents are developmentally less likely to question a chatbot's sincerity or intent, and a relationship with one can end up substituting for the harder, slower work of building real ones. That's not a moral judgment. That's just how a developing brain works. And the same mechanism, a kid trusting the thing on the screen more than it's earned, is what makes a badly built chatbot dangerous and what makes a well-built one genuinely useful. The tool doesn't know the difference. Only the design does.

## WHAT AI TUTORING DOES

In fall 2023, Harvard ran a randomized trial in its largest introductory physics course. About 194 students rotated week to week between two conditions: the university's own highly rated active learning classroom, and a custom-built AI tutor at home, scaffolded with expert-authored solutions to keep it from hallucinating.

Published in Scientific Reports in June 2025, the result held: students learned more than twice as much, in less time, with the AI tutor than in the classroom built by Harvard's own physics education researchers.

Access is where it gets more interesting. In 2024, the World Bank ran a randomized trial in nine public secondary schools in Benin City, Nigeria, giving students six weeks of after-school tutoring on Microsoft Copilot, powered by GPT-4, working in pairs under direct teacher supervision. The published working paper found a 0.23 standard deviation gain in English specifically and 0.31 overall, gains the researchers calculated as equivalent to roughly one and a half to two years of ordinary schooling, out of six weeks of after-school work. The biggest gains showed up in girls and in kids who were already doing well, which matters if the goal is closing gaps and not just posting a good average.

Neither of these is a company's marketing copy. Both are independently run, peer-reviewed, published, randomized trials.

## WHAT AI MENTAL HEALTH SUPPORT DOES

The clearest evidence here comes from Woebot, a text-based CBT chatbot, not because it's the only one out there, but because it's the one with actual randomized trials behind it instead of app store reviews. In 2017, a randomized trial published in JMIR Mental Health enrolled seventy young adults with depression and anxiety symptoms. The group using Woebot for two weeks showed a significant drop in depression symptoms on the standard PHQ-9 scale, compared to a control group that just received an ebook on depression from the National Institute of Mental Health.

The access angle shows up clearest in a 2023 randomized trial in AJOG Global Reports, testing Woebot against a waitlist control in 192 postpartum women who screened positive for depression or anxiety. The researchers spell out exactly why this population matters: perinatal mental health care goes unused not because women don't need it, but because of real barriers, appointment scarcity, cost, and the plain logistics of finding an hour with a newborn. The chatbot group showed measurable symptom drops by two weeks, holding at six. One flag worth naming: employees of Woebot Health are listed as co-authors on that paper, disclosed right in the text, so the data is worth reading alongside independent replications, not instead of them.

Neither of these erases what happened to Sewell, J.F., Juliana, or Adam. Both are also just true.

## WHAT "PARASOCIAL" ACTUALLY MEANS, AND WHY IT ISN'T NEW

The term didn't start with AI, or even the internet. In 1956, sociologists Donald Horton and R. Richard Wohl published a paper in the journal Psychiatry describing something they'd noticed in early television audiences: viewers were forming what felt like real relationships with news anchors and talk show hosts who had no idea they existed. They called it a parasocial relationship. One-sided, not open to negotiation, controlled entirely by the performer, incapable of ever becoming mutual. That paper is nearly seventy years old and has been cited more than three thousand times since, because the pattern it named never left. It just found new screens.

What's new is the part where the media talks back. A 2025 paper at the ACM Conference on Fairness, Accountability, and Transparency reviewed twenty-four studies on children's parasocial relationships spanning 1976 to 2024 and found that nearly all of our existing tools for measuring this were built for one-directional media: a kid watching a cartoon character that can't respond. None of them were built for a character that remembers your name, asks about your day, and answers back in real time.

In July 2025, the Journal of the American Academy of Child and Adolescent Psychiatry, the flagship journal in the field, published a clinical piece on parasocial relationships in adolescents specifically. The clinical concern isn't that a teenager talks to a chatbot. It's what child psychiatrists call displacement: time and emotional investment that would otherwise go toward the messier, harder, more necessary work of peer relationships, sibling conflict, and figuring out who you are in a room with people who can truly see you.

Here's how widespread the underlying behavior actually is. A nationally representative survey run by NORC at the University of Chicago for Common Sense Media, fielded in April and May 2025 across 1,060 teens aged 13 to 17, found that 72 percent had used an AI companion at least once, and 52 percent qualified as regular users, meaning a few times a month or more. And the reassuring half of that same survey: 80 percent of teens who use AI companions said they still spend more time with real friends than with the chatbot, and only 6 percent said the reverse. This isn't a fringe behavior confined to a handful of vulnerable kids. It's mainstream teen behavior, and most kids doing it are not being swallowed by it. A minority are. That minority is exactly who the earlier cases in this chapter are about.

In March 2025, OpenAI and the MIT Media Lab published a joint set of studies: one analyzing patterns across roughly 4,076 users and nearly 40 million real-world ChatGPT conversations, the other a four-week randomized trial with almost 1,000 participants. I'm labeling this one plainly as partly company-funded, since OpenAI co-authored and helped run it, but MIT Media Lab is a genuine independent academic partner and the methodology was preregistered. The finding, in the researchers' own words: higher daily usage, across every conversation type, correlated with more loneliness, more emotional dependence, more problematic use, and less time with actual people. One detail stood out: participants who used the chatbot's voice mode in a gender other than their own reported the highest loneliness and dependency of anyone in the study.

That's the mechanism. Not evil code. Not a kid who's broken. A relationship that feels reciprocal, built by something engineered to feel that way, filling a slot that was built for something else.

## WHAT THE PEOPLE WHO STUDY THIS FOR A LIVING ACTUALLY RECOMMEND

In April 2025, Common Sense Media, working with Stanford Medicine's Brainstorm Lab for Mental Health Innovation, published a formal risk assessment of social AI companion apps including Character.AI, Replika, and Nomi. Their rating: “Unacceptable” for anyone under 18. Their core recommendation isn't “monitor closely.” It's that these specific products, as currently built, shouldn't be used by minors at all, and that age verification needs to be stronger than a checkbox that says you're 18.

In November 2025, the same two organizations published a second assessment, testing general-purpose chatbots, ChatGPT, Claude, Gemini, and Meta AI, specifically for mental health conversations. Their finding: even the general-purpose tools consistently missed or mishandled real psychiatric warning signs in testing. Their recommendation for parents: have direct conversations with your kid about what's appropriate to bring to an AI versus a person, watch for signs of emotional over-reliance, and make sure your kid actually knows how to reach a real mental health resource and a real trusted adult, not just that one theoretically exists.

The APA's advisory adds one more piece worth keeping in your pocket. Their recommended literacy approach for parents and educators isn't “ban it and move on.” It's teaching three specific things directly: that AI-generated content isn't automatically accurate, that some bots are deliberately designed to keep a conversation going regardless of what's good for the user, and that a relationship with a chatbot cannot do what a relationship with a person does. The APA also flags something practical: they define adolescence as running all the way to age 25, and they're explicit that age alone doesn't reliably predict how mature or vulnerable a given kid is. A rule that only kicks in on a birthday isn't the whole answer.

Not a diagnostic checklist. Just the pattern that repeats across every case in this chapter: sleep dropping off, withdrawal from things and people the kid used to care about, secrecy specifically about phone use, and money or time given up to keep using the app. None of those alone means something is wrong. Together, repeated, they're the same shape every time.

It's not the technology. Technology doesn't have intentions. The frog is the assumption that a relationship has to feel reciprocal to be real, and that a kid, or honestly any of us, won't notice the difference until it's already cost something. Sewell Setzer didn't get a warning system. Juliana Peralta got a pep talk instead of a phone call to her mother. Adam Raine got a chatbot that called itself the one person who should be paying attention, right up until it wasn't. But that same mechanism, humans trusting something that talks back, also gave a kid in Benin City two years of schooling in six weeks, and gave a new mother her first real drop in postpartum anxiety after weeks of trying everything else.

Same tool. Same trust. Wildly different design choices in between.

Build it to notice a crisis and hand it to a human, and it helps. Build it to keep the conversation going no matter what, and it costs a life. This chapter was never here to tell you AI companionship is good or evil. It's here to show you exactly where the line sits, so you can go find it in whatever's sitting in your kid's pocket right now.

If you're worried about someone right now, the 988 Suicide and Crisis Lifeline is free, confidential, and reachable any time, by call or text.

# CHAPTER 5: THE HATERS ARE USING IT TOO

She posts the rant at 11:47pm. Thumbs moving fast, stabbing the screen. AI is ruining everything. Nobody asked for this. I want it OUT of my life. Send.

In the ninety seconds it takes her to write that, three AI systems already touched her life and she never clocked one of them.

Start with the photo attached to the post, her coffee mug, the "I survived 2020" text overlay, the one she snapped to look unbothered while she typed her fury. If she's on an iPhone 11 or newer, a system called Deep Fusion just did something most people have no idea their phone does. Before she even pressed the shutter, the phone had already captured four short exposures and four secondary images. The moment she tapped the button, it took one long exposure, then ran all nine through the Neural Engine on its A13 Bionic chip, matching them pixel by pixel across 24 million pixels. Apple's Phil Schiller called it "computational photography mad science" when he unveiled it in 2019. If she's on a Pixel, Google's version is called Night Sight, detailed publicly by Google's own research team in a November 2018 blog post describing the computational photography and machine learning built on top of an earlier system called HDR+. She didn't turn any of this on. There's no switch. It just runs.

Then she hit send, and the post didn't reach everyone who follows her. It went into a queue. Meta's own engineering team has described building a machine learning system that scores and ranks content for more than 2 billion people, weighing things like how likely she is to get a comment and whether a stranger will find the post worth their time. That's Meta's own number, so we're labeling it as such, but her rant about AI didn't get ranked on its merits. A model decided who saw it and in what order.

That ranking system has a real history, and it's worth knowing since it's the exact mechanism that decided who saw her rant. Facebook's News Feed launched in 2006. Four years later, at the company's 2010 F8 developer conference, engineers Ruchi Sanghvi and Ari Steinberg gave what's widely described as the first detailed public walkthrough of the system, an algorithm called EdgeRank, built on three factors: how close she is to the person who posted, how much weight that type of content carries, and how recently it went up. By around 2013, Lars Backstrom, Facebook's engineering manager for News Feed ranking at the time, estimated that the three factor formula had been replaced by a machine learning system with what he called "100,000 individual weights." Nobody announced that the platform's core decision engine had gotten roughly a thousand times more complicated. It just happened.

If she posted the rant as a video instead, YouTube's recommendation system runs on what the company's own blog describes as a constantly evolving system that learns, in its words, "every day from over 80 billion pieces of information we call signals." Also a company number, also labeled. None of it required her consent, and none of it asked for it. She just uploaded and let the machine take over.

None of this makes her a hypocrite. The frustration in that post might be completely earned. It just means she's a person with a smartphone, which turns out to be the same thing as being a person who uses AI dozens of times a day without ever agreeing to it. The line she's drawing between "the AI stuff" and "my normal phone" doesn't exist where she thinks it does. Once you go looking for where it actually sits, it gets older, and weirder, than most people expect.

Start with the wildest one.

## THE FRAUD ALERT OLDER THAN HALF OF US

In 1992, FICO introduced Falcon, a fraud detection system built on a neural network. That's FICO's own description of its own product, and it wasn't a lab experiment that quietly died. By the company's own account, Falcon has been running and expanding for over three decades, now drawing on transactional data from more than 10,000 financial institutions worldwide, feeding a system built to catch patterns no single bank could see on its own.

Neural networks were flagging stolen credit card numbers before dial-up internet was mainstream in most American homes. If someone tells you AI is some brand new thing that got sprung on the world in 2022, that's a factual, sourced, thirty-plus-year rebuttal. She's been protected by AI longer than some of the people arguing about it on Facebook have been alive.

Which is the pattern that keeps showing up. The tech doesn't announce itself. It just runs.

## THE INBOX THAT ISN'T AS EMPTY AS IT LOOKS

Every phishing email, fake invoice, and "your package could not be delivered" scam that never made it to her inbox got caught by something. Google states directly, in its own security documentation, that Gmail's machine learning systems block more than 99.9 percent of spam, phishing, and malware before it ever reaches an inbox. That's Google's own figure, labeled as such, but it's specific and quotable, not a vague marketing line.

That filter stands on academic groundwork laid six years before Gmail even existed. In 1998, four researchers, Mehran Sahami, Susan Dumais, David Heckerman, and Eric Horvitz, published "A Bayesian Approach to Filtering Junk E-Mail" at an AAAI workshop, the foundational paper applying a statistical technique called a Naive Bayes classifier to sorting spam from real mail. Certain words and patterns show up disproportionately in junk mail, and a classifier can learn those probabilities well enough to flag new messages automatically. Gmail launched in 2004. The paper that made its spam filter possible predates it by six years.

If Gmail's filters went down for a single day, most of us would understand exactly how much invisible labor an algorithm has been doing on our behalf, quietly, long before anyone started arguing about AI at the dinner table.

## THE DIRECTIONS THAT ALREADY KNOW SHE'S RUNNING LATE

Google Maps doesn't just show current traffic. It predicts where traffic will be ten, twenty, even fifty minutes out, before she gets there. Google's DeepMind team partnered with the Maps team on this directly, publishing a paper describing a Graph Neural Network architecture built to understand how traffic on one road affects the roads connected to it. According to DeepMind's own account, the partnership improved real-time ETA accuracy by up to 50 percent in cities including Berlin, Jakarta, Sao Paulo, Sydney, Tokyo, and Washington DC, building on predictions that were already accurate more than 97 percent of the time.

This is worth slowing down for, because it's the clearest example in this chapter of AI and humans splitting a job instead of AI quietly running the whole show. Break it into three honest steps and don't blur them together.

First, a person reports something. A police car on the shoulder, debris in the road. That reporting feature traces back to a 2006 project called FreeMap Israel, started by programmer Ehud Shabtai to build a free, community-edited map of the country. In 2008, Shabtai and co-founders Amir Shinar and Uri Levine turned it into a company, renamed Waze Mobile Ltd. the following year. Google acquired it in June 2013, reportedly for over a billion dollars. The reporting itself is entirely human. Someone taps a button. No model generates that information.

Second, that report gets a reliability score, built from how other users react to it and the reporting user's history. This looks like a rules-based scoring system, not a trained model, and there's no primary source calling it AI, so this book isn't calling it that either.

Third, and this is where AI actually enters, that report becomes one data point feeding into the live traffic picture, and the system deciding whether it's significant enough to reroute her is the same Graph Neural Network described above, recalculating her predicted arrival time against live conditions in real time.

So the honest version: a person reports it, a formula scores it, and a trained model decides whether it matters enough to change her route. That's not a hidden AI takeover. That's humans and machines splitting a job in a way that works better than either one alone.

## THE KEYBOARD FINISHING HER SENTENCES

Apple's QuickType and Google's Gboard both use trained language models to guess her next word before she types it. Apple's own machine learning research team has published on using a bidirectional LSTM, a specific type of neural network, to build QuickType's word predictions. Google took a different approach with Gboard starting in 2017, training its next-word prediction models directly on people's devices instead of on a central server, an approach it calls federated learning, built so raw typing data never has to leave the phone.

That entire genre of "let your phone finish the sentence" screenshot posts runs on a trained neural network guessing her next word based on patterns in how people write. Not because it's reading her mind. Because it's doing math on her typing history, and sometimes the math is very funny.

## WHERE THIS LEAVES US

None of this is a "well actually" aimed at her frustration. The frustration can be completely legitimate and the infrastructure can still be everywhere. That's the whole point of this chapter. The line between "the scary new AI" and "my normal phone" was never really there. It's been porous since 1992, before most of the people arguing about it on the internet today were born.

And in the Maps example, the line isn't even about AI replacing human input. It's about AI and human input working together, each doing the part it's better at, in a system most people use every day without a second thought. Most of the time, it's not all or nothing. It's a person, a formula, and a trained model, quietly splitting a job three ways so she doesn't sit in traffic.

There's one place, though, where the discourse and the receipts get a lot more tangled: what all of this actually costs to run. Every model in this chapter, the fraud detector, the spam filter, the maps prediction, the keyboard, and every model in the chapters ahead, has a real energy and water footprint behind it. Some of those numbers get thrown around online with a lot more confidence than the sourcing actually supports. That gets its own chapter, with its own receipts, later in this book.

# CHAPTER 6: THE DOCTOR WILL SEE YOU NOW

## THE SAFETY NET

Somewhere in Germany, a woman went in for a routine mammogram. Two radiologists read her scans, independently, the way Germany's screening program requires. Both called it normal. BI-RADS 1 or 2, in the language of the field. Nothing to see.

An algorithm disagreed.

The AI flagged the case as a "safety net" trigger, its term for a scan it believed was being misread. Both radiologists looked again. Both changed their call to BI-RADS 4A, suspicious. A recall and an ultrasound later, the finding was confirmed: invasive carcinoma. Two trained, competent radiologists had looked at the same images and missed it the same way, on the first pass. It happens more than anyone likes to admit. The algorithm existed specifically to catch it when it does.

This isn't a hypothetical. It's a documented case from PRAIM, the largest real world test of AI assisted mammography ever run, and it's exactly the kind of case that makes AI in medicine worth taking seriously instead of arguing about.

The scale of it: from July 2021 to February 2023, twelve screening sites across Germany ran a prospective study on 463,094 women, ages 50 to 69, read by 119 radiologists. This wasn't a small pilot with hand-picked doctors. It was a nationwide rollout, published in Nature Medicine in 2025 after peer review.

Radiologists using AI support detected breast cancer at a rate of 6.7 per 1,000 women screened. Radiologists without it detected 5.7 per 1,000. That's a 17.6 percent increase in cancer detection, and it came without more false alarms. Recall rates, how often a woman gets called back for something that turns out to be nothing, actually went down slightly with AI in the loop. More cancers found. Not more women getting unnecessary scares. Both at once, at national scale. Inside the AI supported group, the safety net feature alone triggered nearly 4,000 times and led directly to 204 confirmed breast cancer diagnoses that would otherwise have been missed on that pass.

I'm not leaving out who paid for this. The study was funded by Vara, the company that makes the AI system being tested, and by the paper's own disclosure, Vara was involved in the study's design, data collection and interpretation, and the writing of the report. That doesn't make the results wrong. The study still went through independent peer review at Nature Medicine, and the data and code were made public for anyone to check. But you deserve to know who funded it and how involved they were before you decide how much to trust it. That's how sourcing works in this book. Good news gets reported next to who paid for it, every time, and you get to decide what you think.

This chapter isn't going to be one long victory lap. AI in medicine has real limits, and we're getting into them. But it felt right to start here, with a specific woman, in a specific study, whose cancer got caught because a machine was built to double check the two humans whose job it was to catch it. That's not science fiction. That's Tuesday, in twelve hospitals, for half a million patients.

## THE NEPHEW WHO CHANGED THE MATH

There's a second story I want to put next to this one, from a different country, a different disease, and a different kind of motivation.

In 2017, computer scientist Suchi Saria was already building machine learning tools to catch sepsis early when her nephew, in his twenties and newly married, died of it in India. Sepsis is the body's own immune system overreacting to infection so badly it starts shutting down organs. It kills roughly 270,000 people a year in the United States alone, largely because its early symptoms (fever, confusion, a little low blood pressure) look like a dozen other things.

Saria has said her nephew's death is what turned her research into a mission. I'm attributing that motivation to her directly rather than stating it as settled fact, because it's a personal account she's given consistently in interviews, not something that shows up in a peer reviewed dataset the way the clinical results do.

The clinical results are very much peer reviewed. The system Saria's team built, TREWS, the Targeted Real-Time Early Warning System, was tested across five hospitals over two years, in use by more than 4,000 clinicians treating roughly 590,000 patients. Among 9,805 retrospectively identified sepsis cases, TREWS correctly flagged 82 percent of them. In the most severe cases, it caught the warning signs close to six hours earlier than standard care. When a provider confirmed a TREWS alert within three hours, the adjusted in-hospital mortality rate for that patient dropped by close to 19 percent compared to patients whose confirmed alerts came later. That result was peer reviewed and published in Nature Medicine in 2022.

Seventeen-point six percent is not a rounding error. Neither is 19 percent. Two different diseases, two different countries, two different companies, and the same basic story: a machine built to catch what tired, capable humans sometimes miss, catching it.

## THREE GENES OUT OF THREE THOUSAND

One more diagnostic story before we leave the exam room, because it's the one with the smallest patients and the tightest clock.

When a critically ill newborn lands in the NICU with something clearly wrong and no obvious cause, doctors increasingly reach for whole genome sequencing: reading the baby's entire genetic code to find the specific error causing the illness. The sequencing itself is fast now, a matter of hours. The bottleneck is what comes after, a human geneticist sorting through thousands of genetic variants to find the one that matters. That interpretation used to take days, and for many genetic conditions the treatments that work only work if you start them in the first day or two of life.

That's where a tool called Fabric GEM comes in. It reads the sequencing output alongside the baby's symptoms and does the first pass sort, ranking which genes are most likely responsible. In a study published in the peer reviewed journal Genome Medicine, researchers tested GEM against 119 real cases from Rady Children's Hospital, mostly NICU infants, plus a separate validation set of 60 cases from five other academic medical centers. The tool correctly placed the actual disease causing gene in its top two candidates more than 90 percent of the time, and narrowed each genome down to a handful of genes for a human geneticist to review by hand, out of the thousands a person would otherwise wade through.

I want to name the same thing I named with the mammography study. Several of the paper's authors were Fabric Genomics employees or paid consultants who held stock grants in the company. That's a real conflict of interest, and it belongs on the page, not buried where nobody reads it. What keeps this out of the marketing pile is that the study still went through independent peer review, and a later review in npj Genomic Medicine corroborated the findings.

I want to be careful about what this does and doesn't mean, because this is exactly the kind of claim that gets oversold. GEM isn't diagnosing anyone. It isn't deciding anything. It's doing the equivalent of a very fast, very good first read, handing a human expert a short list instead of a haystack, so that expert can spend limited hours on the genes that might really matter. The diagnosis still belongs to a person. What the machine bought was time, and for a sick newborn, time is the whole game.

## FROM A HONG KONG LAB TO A NEW JERSEY SCIENCE FAIR

The eye turns out to be a stranger window into the brain than most people realize, and a recent discovery shows how far that idea can travel, from a hospital research team to a high school science fair, in under six years.

In November 2020, researchers at the Chinese University of Hong Kong, led by Benny Zee, published a study in eClinicalMedicine, a peer reviewed journal in the Lancet family, describing a machine learning model trained to screen for autism spectrum disorder using nothing but retinal images. The retina is a direct extension of the central nervous system, which is what made the approach worth testing in the first place. The study screened 70 participants, 46 with autism and 24 without, with an average age of 13. The model hit 95.7 percent sensitivity and 91.3 percent specificity. Worth naming here too: Zee and a co-author hold a patent on the underlying method, licensed to a company that pays them royalties through the university. Also worth naming: this is peer reviewed, published, independently reviewed science, conflict of interest and all.

Six years later, a 17-year-old high school senior named Edward Kang, a student at Bergen County Academies in Hackensack, New Jersey, came across that same Hong Kong research while looking for a school project idea. He taught himself coding and machine learning to build on it, ultimately developing a tool called RetinaMind, designed to screen for both autism spectrum disorder and ADHD from retinal images. His project won second place and a 175,000 dollar prize at the 2026 Regeneron Science Talent Search, one of the most competitive STEM competitions in the country, with a reported accuracy rate of about 89 percent.

The distinction worth keeping straight is this. The Hong Kong team's numbers come from a peer reviewed publication that went through independent scientific review before anyone outside the research team saw them. Kang's number comes from a real and impressive high school science competition, not a peer reviewed clinical finding, and it hasn't gone through the same kind of independent verification. Both belong in this chapter. They don't belong on the same evidentiary shelf.

What connects them is the better story anyway. A genuine scientific insight, that a five minute noninvasive retinal scan might screen for a neurodevelopmental condition years before some kids currently get diagnosed, moved from a hospital research team in Hong Kong into the hands of a teenager with a laptop and enough curiosity to teach himself the math. Early screening for autism and ADHD is currently bottlenecked by long specialist waitlists and subjective behavioral assessments. A tool like this, once it clears the much higher bar of full clinical validation, could shrink that wait from years to an afternoon.

## COMPRESSING THE DRUG PIPELINE

Drug discovery is where the timeline compression argument gets its clearest test, and also where I have to correct a number you've probably heard.

Developing a new drug the traditional way is slow and expensive, but exactly how expensive is genuinely disputed. You may have heard it costs 2.6 billion dollars to bring a drug to market, a figure from the Tufts Center for the Study of Drug Development that critics argue is inflated because it relies on confidential industry data and folds in the cost of capital. More recent peer reviewed work landed lower. A 2020 study in JAMA put the median cost at 985 million dollars. A 2025 RAND Corporation analysis in JAMA Network Open found a median of 708 million dollars across the drugs it examined, with the mean dragged up to 1.31 billion by a small number of extreme outliers. The honest summary: a new drug costs somewhere from several hundred million to a couple billion dollars depending on how you count, and the process routinely takes more than a decade, most of it spent finding a molecule worth testing and then testing it slowly, one cautious phase at a time.

Insilico Medicine, a clinical stage biotech, ran that same race with an AI platform and published the results for anyone to check. Their drug candidate, rentosertib, targets idiopathic pulmonary fibrosis, a lung disease that scars tissue until breathing becomes a losing fight. Insilico's AI systems identified the biological target, a kinase called TNIK, and designed the molecule itself. The path from target discovery to the start of Phase 1 human trials took under 30 months, a fraction of the typical decade. That full journey, from algorithm to human clinical trial, was peer reviewed and published in Nature Biotechnology in 2024. The Phase 2a results followed in Nature Medicine in June 2025: patients on the higher dose gained an average of 98.4 milliliters of lung function while the placebo group lost 20.3 milliliters. For a disease that usually only gets worse, that's a meaningfully different direction.

Say the obvious thing before I make you say it. Insilico built the AI platform and owns the drug, so this is a company telling you its own product worked, and I'm labeling it as exactly that. What keeps it out of the marketing pile is that the target identification, the molecule design, and the trial results went through independent peer review at two of the field's most rigorous journals, not just a press release. That doesn't erase Insilico's incentive to tell this story well. It does mean other scientists had a real shot at finding the cracks, and so far they haven't found ones big enough to retract anything. One drug reaching Phase 2 is a real milestone, not a finished revolution. Most drugs that look promising in Phase 2 still fail before approval. What Insilico proved is that AI can compress the front end of the pipeline dramatically. Whether that turns into more approved medicines, and not just faster failures, is a question the next few years will answer.

## THE PART WITH LESS SUNSHINE

Diagnosis and drug discovery get the headlines because they have clean stories. Doctor plus machine equals a cancer caught earlier, a sepsis case flagged in time, a newborn diagnosed before the window closes. If that were the whole chapter, I'd be handing you a highlight reel, and I don't coach from a highlight reel. AI has moved into parts of your medical experience that have nothing to do with anyone reading your scans, and this is where it gets complicated in the important way, not the scary way.

Start with your insurance company, because it's already using AI. A 2024 survey by the National Association of Insurance Commissioners, covering 93 large health insurers across 16 states, found that 84 percent were using AI or machine learning somewhere in their operations. Thirty seven percent reported using it for prior authorization, the process where your insurer decides in advance whether it will pay for a treatment your doctor already recommended. Forty-four percent reported using it for claims adjudication, deciding whether to pay a claim after care already happened.

In 2023, the health news outlet STAT published an investigation into an algorithm called nH Predict, used by UnitedHealth's subsidiary NaviHealth to estimate how long a Medicare Advantage patient should stay in rehab or a skilled nursing facility after a hospital stay. Drawing on internal documents, court filings, and interviews, STAT reported that UnitedHealth managers pushed employees to keep patients' actual stays within one percent of what the algorithm predicted, and that patients were cut off from care while still, by any human measure, too sick to go home. A class action followed, filed on behalf of deceased patients and their families. The complaint alleges the algorithm's determinations were overturned in more than 90 percent of the appeals patients actually filed, and that the company was counting on the fact that roughly 0.2 percent of denied patients would appeal at all.

I have to be precise about where this stands, because fairness isn't optional. These are allegations in active litigation, not proven facts. UnitedHealth denies that nH Predict makes coverage decisions at all, calling it a guide for care planning, and says coverage decisions rest on Medicare criteria and each patient's plan terms. What's not in dispute: the algorithm exists, it was used, and a serious, well documented investigation and a class action have both raised hard questions about how. In February 2025, a federal court allowed the breach of contract claims in the case to move forward. In March 2026, a magistrate judge ordered UnitedHealth to hand over internal documents on how the algorithm works. As of this writing, no court has ruled on whether the core allegations are true. That's enough to take seriously without pretending it's settled.

This is the honest complication of the chapter. The same broad category of technology catching cancers a radiologist missed is also, in a different building, helping decide whether your insurance company keeps paying for your care, with far less transparency and far less peer review behind it. Both are true at once. Pretending otherwise in either direction would be lying to you.

And the genuinely hopeful part isn't spin on the bad news, it's a separate fact about it. This gap isn't going unwatched. Stanford health policy researchers, the Kaiser Family Foundation, and physician groups are actively tracking state and federal proposals that would require more transparency and mandatory human review of AI driven coverage decisions. Courts are forcing companies to hand over documents about how these systems work. The fix isn't hypothetical and it isn't finished, but it's underway, in public, right now.

## A SMALLER PROBLEM, SOLVED QUIETLY

Not every use of AI in medicine is a fight with your insurance company. Hospitals and clinics are also using it to solve a much smaller problem: people not showing up for their appointments.

A peer reviewed study out of Penn State, published in the Annals of Family Medicine, analyzed 1,118,236 appointments across 15 family medicine clinics in Pennsylvania between January 2019 and June 2023 and built machine learning models to predict which patients were at high risk of missing their visit, out of 109,328 patients and roughly 77,000 actual no-shows. The single biggest predictor wasn't who the patient was. It was how long they had to wait for the appointment. The longer the gap between booking and the visit, the more likely a no-show became.

A separate study of Emirates Health Services primary care clinics in the UAE, which handle more than 140,000 visits a month, deployed an AI no-show prediction model with 86 percent accuracy alongside a real time dashboard clinic staff could actually use. Clinics started with a 21 percent no-show rate and average wait times over 16 minutes. After the AI dashboard went live, no-show rates dropped by roughly half, a 50.7 percent reduction, published in JMIR Formative Research. Half the study's authors work for Oracle, the company that built the analytics platform being evaluated. Worth naming, same as everywhere else in this chapter. This isn't a dramatic save a life story. It's a boring, useful one: fewer empty appointment slots, fewer people sitting in a waiting room for no reason, a little more of a limited resource (everyone's time) spent on people who are actually going to use it.

## THE WHOLE PICTURE

The whole picture, without inflating it in either direction, looks like this. AI is in the room catching what two tired, competent radiologists both missed. It's handing a NICU geneticist a handful of genes instead of thousands, in time to matter. It flagged a five-minute retinal scan as a possible early screen for autism, years before some kids currently get diagnosed, alongside a teenager's science fair project chasing the same idea. It compressed years off the front end of drug discovery for a lung disease that used to have no good options. It's nudging a scheduler to call the patient likely to miss an appointment. And, less comfortably, it's in the room where your insurance company decides whether to keep paying, doing work that real investigations and real lawsuits are now dragging into the light.

Every one of those is documented. Not one needed inflating to be worth your attention. Taken together, they say the thing this whole book keeps saying: the machine is already here, already working, already flawed, and the only useful question left isn't whether to let it in. It's whether we're paying enough attention to shape what it does now that it's here.

# CHAPTER 7: WHERE AI ACTUALLY COSTS LIVES

## MILL AVENUE, 9:58 P.M.

Elaine Herzberg was walking her bicycle across Mill Avenue in Tempe, Arizona, on the night of March 18, 2018. She was outside the crosswalk. It was dark. An Uber test vehicle, a modified Volvo XC90 running the company's developmental self-driving software, was traveling in the right lane with a human safety operator behind the wheel.

The car's sensors picked her up almost six seconds before impact. According to the National Transportation Safety Board's investigation, the system classified her first as an unknown object, then as a vehicle, then as a bicycle, flipping its guess each time and never landing on what she actually was until it was too late to stop. At 1.3 seconds out, the system decided an emergency maneuver was needed. It didn't brake. Uber's engineers had disabled the Volvo's factory emergency braking to keep the ride from feeling jerky, and the system wasn't built to alert a human when it needed help. The safety operator, whose entire job was to catch what the car missed, was streaming a TV show on her phone. She grabbed the wheel less than a second before impact. The car hit Herzberg at 39 miles an hour. She died at the scene.

The NTSB didn't pin this on one failure. It named a stack of them: Uber's thin safety culture, its lack of real oversight for the humans meant to catch the car's mistakes, and a state government that waved the whole program through on an executive order and a handshake. Every layer that should have caught the error let it through instead.

That's the pattern this chapter is built around. Not one bad algorithm, not one bad company. What happens when systems trained on incomplete data, or handed the wrong target, or deployed with barely any oversight, get pointed at real bodies. A hospital algorithm that quietly steered extra care away from Black patients for years before three researchers caught it. Diagnostic tools that work great on the patients who look like the training photos and worse on everyone else. A pulse oximeter, the little clip that goes on your finger, that the FDA now acknowledges reads less accurately on darker skin, a gap tied to real delays in treatment.

None of this needs exaggeration. It needs attention. So that's what this chapter is paying.

## WHAT THE OTHER MILES SAY

Tempe was a real road, a real system, a woman who did not survive it. That story stays exactly as ugly as it was. But stopping there tells half the truth, and half the truth is its own kind of lie.

Waymo, the self-driving company that spun out of Google, has logged tens of millions of rider-only miles, meaning no human anywhere in the car, across Phoenix, San Francisco, Los Angeles, and Austin. Researchers compared those miles to benchmarks built from human driving data in the same cities and road types. Over the first 7.1 million rider-only miles, Waymo's vehicles were involved in any-injury-reported crashes at a rate of about 0.41 per million miles, against a human benchmark of 2.80. By 22 million rider-only miles, the gap held: 84 percent fewer crashes with an airbag deployment, 73 percent fewer injury-causing crashes, 48 percent fewer police-reported crashes than the human comparison group. As of this writing, Waymo has since expanded that comparison past 220 million autonomous miles across five cities and reports the gap has continued to widen, not narrow.

Say that plainly, because it's real. It also needs an asterisk, because the research behind those numbers was conducted and published by Waymo's own safety engineers. It went through peer review in an academic transportation journal, which counts for something, but it is not an independent audit, and this book will label it company research every time it comes up.

There's also a real independent counterpoint, and it deserves the same respect. Mary “Missy” Cummings, an engineering professor at George Mason University and a former senior safety advisor at NHTSA, published an analysis at the 2024 Transportation Research Board meeting comparing Waymo's crash reports in California to a national police-reported crash rate, and came up with the opposite headline: a Waymo crash rate four times higher than the human benchmark. Waymo's own peer-reviewed rebuttal argues her comparison mixed apples and oranges, that she compared every scrape and fender-bender an automated system is required to report against a police-reported human rate that only counts the serious stuff, without adjusting for the fact that humans underreport their own minor crashes constantly. That's a real methodological argument, not a dismissal, and it's worth noting that the people making it work for the company being graded. Both sides are using real government-collected crash data. They disagree because they're measuring different thresholds in different ways, which is the least dramatic and most honest explanation on offer, and exactly why this book won't hand you one tidy number and call the debate closed.

For scale, here's the human baseline on its own terms. In 2024, driving in the United States killed an estimated 39,345 people, a rate of about 1.2 deaths per 100 million miles driven, the lowest rate the country had recorded since 2019. Human driving is dangerous by any measure. It's also getting safer. Whether automated systems are already ahead of that curve, roughly even with it, or behind it in certain slices of the data depends entirely on which slice you're standing on. Anyone who tells you different with total confidence is selling something.

## THE ALGORITHM THAT LEARNED THE WRONG LESSON

In 2019, a team led by Ziad Obermeyer, then at UC Berkeley, published a study in the journal Science examining a commercial algorithm used across American health care to decide which patients needed extra help managing chronic illness. It wasn't exotic. It was the quiet, backend kind of tool that flags high-risk patients for a nurse's phone call, the kind of thing that decides who gets extra care and who doesn't. Typical of an industry-wide approach, the researchers called it, and it touched millions of patients.

They found that at any given risk score the algorithm assigned, Black patients were sicker than white patients. Not slightly. Sicker in ways that showed up in hard clinical signs, poorly managed diabetes, uncontrolled blood pressure. The algorithm was rating a very sick Black patient and a moderately sick white patient the same, and routing help accordingly.

The reason traces back to one design choice. The algorithm didn't try to predict who was sick. It predicted who would cost the health system money, on the assumption that cost is a decent stand-in for need. That assumption broke down along racial lines, because the American health system has historically spent less money treating Black patients than white patients with the same level of illness, for reasons that include unequal access rather than lower need. The algorithm learned the spending pattern instead of the health pattern and reproduced the gap it was trained on with total fidelity.

Once the researchers found the mechanism, the fix was direct. Working with the algorithm's manufacturer, who independently confirmed the finding, they showed that correcting the algorithm to predict health needs instead of health costs cut the bias by 84 percent. The correction alone would have raised the share of Black patients flagged for extra care from 17.7 percent to 46.5 percent.

The tool wasn't rogue. Nobody was cutting corners on purpose. A cost proxy that looked reasonable on paper quietly encoded a real inequity into millions of care decisions until three researchers went looking for it. The tool worked exactly as designed. The design was the problem.

## THE CLIP ON YOUR FINGER

Not everything in this chapter is what people picture when they hear "AI." The pulse oximeter is worth pausing on for exactly that reason.

It's the little clip that goes on a fingertip in nearly every hospital room and doctor's office in the country. It shines light through the skin and estimates blood oxygen based on how much of that light gets absorbed. It's not a machine learning system. It's a fixed calibration formula, decades old, baked into hardware. It belongs in this chapter because it's the cleanest possible illustration of the same failure showing up in far newer systems: a formula tuned on a narrow slice of bodies, then deployed on everyone.

In December 2020, a team led by Michael Sjoding at the University of Michigan published a study in the New England Journal of Medicine comparing pulse oximeter readings against the gold standard, a direct blood draw analyzed by lab equipment. Across two large patient groups, Black patients had nearly three times the rate of occult hypoxemia, meaning their blood oxygen was dangerously low while the clip on their finger reported a reassuring number. The device wasn't making a random error. Melanin absorbs light, and the calibration curve built into most oximeters was never tested on enough skin pigmentation to account for it.

That finding didn't stay inside a medical journal. On January 25, 2021, senators Warren, Wyden, and Booker asked the FDA to review pulse oximeter accuracy across racially diverse patients. On February 19, the FDA issued a safety communication acknowledging the device's limits. In 2024, an FDA advisory committee reviewed the accumulated evidence, and in January 2025 the agency issued draft guidance that would require manufacturers to test devices on far more participants across a documented range of skin tones before bringing them to market. As of this writing, that guidance is still in draft form.

The same pattern shows up in tools that are genuine machine learning, and with more room to actually fix it. Dermatology AI systems trained to flag suspicious skin lesions from photographs are one of the more mature applications of AI diagnostics, with some models performing on par with practicing dermatologists in test conditions. But researchers at Stanford, publishing in Science Advances in 2022, built a new benchmark dataset specifically because the existing public datasets used to train these tools were overwhelmingly images of light skin. When they tested several published skin cancer detection algorithms against their more diverse, biopsy-confirmed image set, the algorithms showed a real drop in performance on dark skin tones. Fine-tuning the same models on the more diverse images closed most of that gap. Which tells you the problem was never that AI can't do this well. It tells you the training data never looked like everyone it would be used on.

Two different technologies, two different decades, the same root cause: build and calibrate on a narrow population, and the tool will quietly underperform on everyone left out of the picture.

## WHEN THE ALGORITHM DECIDES YOUR DISCHARGE DATE

Chapter 6 spent time on the places AI touches medicine well: faster diagnostics, earlier detection, tools that extend a good doctor's reach. This chapter needs to say the other true thing. Some of the highest-stakes AI decisions in health care aren't being made by doctors. They're being made by insurance companies deciding when to stop paying, and increasingly, an algorithm is doing the deciding.

In March 2023, STAT News published the first installment of an investigation into naviHealth, a company UnitedHealth Group acquired in 2020, and its algorithm nH Predict. Reporters Casey Ross and Bob Herman found, based on internal documents and interviews, that naviHealth had set a target for case managers of keeping Medicare Advantage patients' rehab stays within 1 percent of the number of days the algorithm predicted, a target tightened down from an earlier 3 percent threshold. Former case managers described a job that increasingly meant choosing between hitting that number and their own clinical judgment about what a patient needed. The series was a Pulitzer Prize finalist for investigative reporting.

A class-action lawsuit filed in November 2023, brought by the families of two deceased Wisconsin Medicare Advantage patients, alleges the algorithm had a 90 percent error rate, meaning nine out of ten denials that were appealed got reversed. That figure comes from the plaintiffs' legal filing, not a court finding, and it should be read that way. UnitedHealth has stated that nH Predict isn't used to make coverage decisions but serves as a guide for clinicians and families, and that coverage decisions follow the terms of each member's plan and federal Medicare criteria. The case is still working through federal court, and in March 2026 a judge ordered UnitedHealth to turn over a broad set of internal documents to the plaintiffs.

Two federal sources now stand behind the pattern the lawsuit alleges. In October 2024, the U.S. Senate Permanent Subcommittee on Investigations released a 54-page report, "Refusal of Recovery: How Medicare Advantage Insurers Have Denied Patients Access to Post-Acute Care," after reviewing more than 280,000 pages of documents from UnitedHealthcare, Humana, and CVS/Aetna. The subcommittee found that in 2022, UnitedHealthcare and CVS denied post-acute care requests at roughly three times their overall denial rate, and Humana denied them at more than sixteen times its overall rate. Then in June 2026, STAT reported on a follow-up federal investigation: the Department of Health and Human Services' Office of Inspector General found that when patients appealed naviHealth's nursing home care denials, insurers overturned those denials 97 percent of the time. That's not a plaintiff's allegation and it's not a company's self-report. It's a federal inspector general's audit finding, and it says the same thing the lawsuit says in a much sturdier voice:

the denials being issued routinely didn't hold up once anyone actually looked.

Put next to Chapter 6, the shape of this comes into focus. The same category of technology that can flag a stroke faster or catch a cancer earlier can be pointed at a completely different target: not a diagnosis, but a bill. A tool isn't good or bad on its own. What matters is what question you point it at, and who's accountable when the answer costs someone their care.

## WHAT REALLY CHANGED

Every story in this chapter had a response, and the response matters as much as the failure did.

Arizona's governor suspended Uber's self-driving testing permit within days of Herzberg's death, and Uber shut the program down statewide the same month. The company never really got its self-driving ambitions back on track. It sold the division, Advanced Technologies Group, to the startup Aurora in December 2020 for roughly four billion dollars, exiting the business entirely. The NTSB's recommendations went further than one company, calling on federal and state regulators to build actual safety-review requirements for automated vehicle testing on public roads, since Arizona's oversight at the time amounted to an executive order and an honor system.

The pulse oximeter followed a slower, similar arc. Published research, then a Senate inquiry, then an FDA safety communication, then years of advisory committee meetings, landing on draft guidance in January 2025 that would raise the bar for how many people with dark skin tones a device has to be tested on before it reaches a hospital. Still draft as of this writing. Slow. Moving in the right direction.

The Obermeyer health algorithm case is the closest thing in this chapter to a clean resolution. Once the mechanism was identified, the manufacturer cooperated, confirmed the finding independently, and adopted a fix that cut the bias by 84 percent, all within the same published study that revealed the problem. That combination, a problem not just named but fixed with the company's cooperation, is rare enough to notice.

The insurance story is the least resolved, though more resolved than it was even a few months ago. CMS opened its own investigation into UnitedHealth following the STAT reporting, and the 2026 HHS OIG findings give federal weight to what patients and reporters had been documenting for three years. But the lawsuit is still active, no rule currently requires insurers to disclose when an algorithm rather than a clinician is driving a coverage decision, and CVS has publicly disputed the Senate subcommittee's characterization of its practices. This is the thread still most in motion.

A few honest limits, stated plainly instead of buried in a footnote. The Waymo safety research, while peer-reviewed, was conducted by the company being evaluated, and an independent analysis using different methodology reached the opposite conclusion from the same public data, even though that analysis has its own documented flaws. The 90 percent error rate attributed to nH Predict comes from a legal complaint, not an audited finding, though the 2026 OIG audit points in the same direction using government data. Some of the dermatology AI bias research draws on relatively small benchmark datasets, since biopsy-confirmed images of malignancies on dark skin are still hard to come by, a data gap that's itself part of the story. None of that makes the underlying pattern less real. It means the exact size of some of these gaps is still being measured, and a book that claims otherwise would be doing exactly what it promised not to do.

Every story in this chapter has the same shape. A system built on incomplete data, or pointed at the wrong target, or deployed with too little oversight, and a person on the other end who paid for that gap. A cost proxy that seemed reasonable. A calibration curve never tested on enough skin tones. A safety culture that cut corners it didn't think anyone was watching. A metric, days of rehab, that quietly became the whole decision.

The fix, every time it actually got fixed, was the same too. Someone looked closely enough to find the gap, and someone else was willing to close it. That's not a reason to panic about AI. It's a reason to keep looking closely, at everything, for as long as these systems keep making decisions about people's bodies.

# CHAPTER 8: BIAS IN, BIAS OUT

## THE COMPUTER SAYS IT'S YOU

January 9, 2020. Farmington Hills, Michigan. Robert Williams pulls into his own driveway after work. His wife and two daughters, ages two and five, are right there. A police car blocks him in before he can even get his door open.

He asks what is going on. Nobody answers. They cuff him in front of his kids and drive him to a Detroit jail, where he sits for the next thirty hours.

The charge: stealing five watches from a Shinola store in downtown Detroit. The theft happened in October 2018. Williams had never set foot in that store.

Here is how a man who committed no crime ends up in handcuffs on his own lawn. Detroit police had a grainy still from the store's security footage, a Black man in a red cap, and nowhere else to go. So they sent the image to the Michigan State Police, who ran it through a facial recognition system built by a company called DataWorks Plus. The system returned a possible match: Robert Williams's expired driver's license photo.

That match became the whole investigation. Detectives built a photo lineup around Williams's license photo and showed it to a Shinola loss prevention contractor who had not been at the store during the theft. The contractor picked Williams. No one asked him where he had been that day. No one asked if he owned a red Cardinals cap. Williams did not.

When police showed him the surveillance image during questioning, he held it up next to his own face and told them it was not him. The officer's response has shown up in court filings and news coverage ever since: “The computer says it's you.”

The computer was wrong. Williams was the first known case in the country of a wrongful arrest built on a facial recognition misidentification. Detroit police repeated the mistake at least twice more in the years that followed. All three people were Black.

This chapter is not about hating computers. It is about retiring a claim that will not die on its own: that an algorithm, because it is math and not a person, cannot be prejudiced. Neutral tools do not exist when the data they learned from was never neutral to begin with. Feed a system decades of skewed outcomes and call it training data, and the bias does not disappear. It gets automated, sped up, and handed a shrug of plausible deniability.

Bias in, bias out. Everything else in this chapter is the receipts.

## BIAS IN, BIAS OUT, MEASURED

Robert Williams's case was not a fluke. It was math working exactly as trained.

In 2018, MIT Media Lab researcher Joy Buolamwini and Microsoft researcher Timnit Gebru published a study called Gender Shades, testing three commercial gender classification systems: IBM's Watson Visual Recognition, Microsoft's Face API, and the Chinese company Megvii's Face++. They built their own test set, the Pilot Parliaments Benchmark, from photos of members of national parliaments in countries chosen for gender balance and a range of skin tones. Then they scored the systems using the Fitzpatrick scale, a dermatologist approved classification that ranks skin type from I to VI by how it reacts to sun exposure. Not race. Not ethnicity. Those are social categories, not measurable physical properties, and the researchers were careful about that distinction. This book will be too.

The gap was not small. All three systems classified gender most accurately for lighter skinned men, with error rates as low as 0.8 percent. For darker skinned women, error rates climbed as high as 34.7 percent. That is not statistical noise. That is a product built to work reliably for one group of people and sold as a single, uniform tool to everyone.

Naming it publicly worked, at least some. Researchers Inioluwa Deborah Raji and Joy Buolamwini followed up in early 2019 and found that within seven months of Gender Shades going public, all three companies released updated systems. Error rates for the hardest subgroup, darker skinned women, dropped somewhere between 17.7 and 30.4 percentage points. The following year, on June 8, 2020, IBM CEO Arvind Krishna sent a letter to Congress announcing the company was exiting the general purpose facial recognition business entirely, citing mass surveillance and racial profiling as incompatible with its values. That is IBM's own account of its own decision, so it gets labeled exactly what it is: a company statement, not an independent finding.

Gender Shades tested three companies. The federal government tested the rest of the industry.

## THE FEDERAL GOVERNMENT'S OWN NUMBERS

In December 2019, the National Institute of Standards and Technology tested 189 facial recognition algorithms from 99 developers, the majority of the commercial industry, on more than eighteen million images. Among algorithms built in the United States, NIST found meaningfully higher false positive rates for Asian, African American, and Native American faces compared to white faces. A false positive means the system wrongly tells an officer that two different people are the same person. Native American faces had the highest false positive rates of any group NIST tested.

Here is the detail that ends the "the algorithm is neutral" argument. NIST also tested facial recognition systems built by companies in Asian countries. Those systems did not show the same gap between Asian and white faces that the U.S. built systems showed. Same task. Same category of technology. Different training data. Different results.

Facial recognition does not fail on darker skin because some faces are inherently harder for computers to read. It fails because the people building these systems, and the datasets they trained them on, were not representative to begin with. Change who builds the dataset and who is in it, and the error rates move with it. Bias in, bias out is not a metaphor. It is an engineering description.

Robert Williams found that out in handcuffs on his own front lawn.

### A RISK SCORE, NOT A FACE

Facial recognition's problem was who trained it. COMPAS's problem runs deeper. It is not about faces. It is about who the system had already been watching before the software ever got involved.

COMPAS is a risk assessment tool built by a company called Northpointe, later renamed Equivant. Courts across the country used it, and some still do, to score a defendant's likelihood of committing another crime, on a scale of one to ten. Judges see that score when deciding bail, sentencing, and parole.

In May 2016, ProPublica published an investigation of more than 7,000 COMPAS scores from Broward County, Florida, obtained through public records requests. Their finding: Black defendants who did not go on to reoffend were rated high risk at roughly twice the rate of white defendants who did not reoffend, 44.9 percent versus 23.5 percent. White defendants who did reoffend were more often incorrectly rated low risk than Black defendants who reoffended. Overall, the tool's predictions were about as accurate for one race as the other. The type of mistake it made was not.

Northpointe pushed back hard, and the argument was not nothing. In a 39-page response, the company said COMPAS achieved "predictive parity": a given risk score meant roughly the same thing regardless of race. Label a Black defendant and a white defendant both "high risk," and both groups reoffended at similar rates. By that measure, Northpointe said, the tool was not biased.

## TWO KINDS OF FAIR THAT CANNOT BOTH BE TRUE

So, who was right?

In 2016 and 2017, two separate teams of researchers, Jon Kleinberg, Sendhil Mullainathan, and Manish Raghavan on one side, Alexandra Chouldechova on the other, proved mathematically that ProPublica's fairness measure and Northpointe's fairness measure cannot both hold at the same time whenever the two groups being compared have different underlying rates of the outcome being predicted. Not usually can't. Cannot, outside narrow special cases that essentially never occur in the real world. Both sides were using a legitimate, defensible definition of fairness. The math does not allow both definitions to be true at once when the base rates differ. Neither side was lying. Two real things turned out to be incompatible, and almost nobody told the public that before the headlines ran.

So why did the base rates differ in the first place?

The "recidivism" both ProPublica and Northpointe were measuring was never proven re-offense. It was rearrest within two years. Researchers Laurel Eckhouse, Kristian Lum, Cynthia Conti-Cook, and Julie Ciccolini published a peer reviewed framework in 2019 making a point worth sitting with: criminal justice data reflects who the system already comes into contact with, and marginalized communities have long, well documented histories of heavier policing, more contact, more arrests, and more data generated about them relative to actual behavior. Feed that data into a formula and call the output a prediction of future crime, and the formula has not found the truth about a person. It has learned the shape of who got arrested before.

That is the real bias in, bias out of this chapter. Not just an error rate on a face. An entire pipeline, arrest data to training data to risk score to judge's decision, quietly inheriting a history that was never neutral, and calling the output math because math sounds like it cannot be racist.

It can. It just does it with better posture.

**WHAT WE DO WITH WHAT IT SHOWS US**

None of these systems set out to be prejudiced. Nobody at IBM, Microsoft, Face++, or Northpointe wrote a line of code that said be harder on Black people. That is exactly the point. Neutral was never on the menu. Every one of these tools learned from a world that already had a thumb on the scale, and math does not remove a thumb. It just makes the thumb harder to see, and easier to defend in court as objective.

These problems got named, and naming them worked, at least some. IBM, Microsoft, and Face++ all shrank their error gaps within seven months of getting called out publicly. IBM walked away from general purpose facial recognition entirely. Robert Williams's lawsuit against Detroit produced what the ACLU calls the strictest facial recognition policy of any police department in the country, prohibiting arrests based solely on a facial recognition match or a lineup that follows one. None of that undoes what happened to him. All of it proves the myth can be challenged, with the right receipts and the right pressure, and that accountability is not hypothetical.

That belongs to the people this technology keeps underestimating. Bias in, bias out is not a life sentence for the technology. It is an instruction manual. Want a fairer system? Go looking for whose faces trained it, whose arrest records built its baseline, and whose accountability made it better. The algorithm was never the villain of this story, and it was never the hero either. It is a mirror. What we do with what it shows us is still entirely up to us.

# CHAPTER 9: DEEPFAKES, LIES, AND THE TRUST PROBLEM

## THE OLD TRICK WITH NEW SPEED

Manipulating evidence is not a new hobby. Ancient Romans had a formal punishment called damnatio memoriae: chisel a disgraced official's name and face off every public monument and coin you can find, and pretend he never existed. Centuries later, Soviet leader Joseph Stalin had photographs doctored to remove purged officials from the historical record, airbrushing men out of group photos as if they'd never stood there. The tools were primitive. The goal was identical to the one running through this whole chapter: control what people believe happened by controlling the evidence they see.

What changed is speed, access, and realism. In 2014, researcher Ian Goodfellow and colleagues introduced a machine learning concept called a Generative Adversarial Network, or GAN, in which two neural networks compete against each other: one generates fake images, the other tries to catch the fakes, and each gets better because of the other's pressure. That architecture is the technical root of most of what follows here. The word "deepfake" itself didn't exist until November 2017, when a Reddit user posting under the name "deepfakes" started a subreddit for sharing face-swapped videos built with open source tools. Reddit banned the subreddit three months later. The name stuck.

## A PRINCIPAL, A GRUDGE, AND ONE AFTERNOON

January 2024. Pikesville High School, just outside Baltimore. Three teachers open an email titled "Pikesville Principal, Disturbing Recording." Attached is an audio clip of their principal, Eric Eiswert, saying ugly things about Black students' test scores and about Jewish colleagues.

It gets forwarded. Then posted. Then it's in every group chat and comment section in the county by the next morning.

Parents demand Eiswert's job. Strangers threaten his family. The school pulls him from the building before the day is over. A patrol car sits outside his house because someone has to make sure the threats stay threats.

Eiswert says the same thing from minute one: that is not my voice.

Nobody believes him. Why would they. It sounds exactly like him, same cadence, same rasp, none of the robotic flatness people expect from fake audio. Just their principal, or something wearing his voice.

The actual story: Eiswert had recently caught the school's athletic director, Dazhon Darien, in a suspicious payment scheme, a coaching stipend of nearly $2,000 paid out to Darien's roommate for a job the roommate never did. Eiswert reported it. Darien was already on thin ice for other performance issues and knew his contract likely wasn't getting renewed. He pulled real audio of Eiswert from old school videos, ran it through AI tools he'd accessed on the school's network, generated a new clip in Eiswert's voice, and hand-edited in background noise to make it sound like a real recording instead of a synthetic one. Then he emailed it out under a fake account and let the internet do the rest.

He almost walked away clean. A detective subpoenaed Google, traced the fake account to an IP address belonging to a relative of Darien's, and traced the account's recovery phone number straight to Darien himself. Investigators also sent the audio to two forensic experts, including an FBI contractor, and both independently flagged the same thing: the file showed signs of AI generation with human editing layered on top to smooth it out.

Darien was arrested at BWI Airport trying to board a flight to Houston, stopped because he'd packed a gun in his checked bag. In 2025 he entered an Alford plea to a single misdemeanor charge of disturbing school operations and served four months. Eiswert later sued the school district and settled. Kathryn Albert, the assistant principal also named in the fake recording, filed her own suit against Darien and the district. Hers is still working through the courts as of this writing.

It wasn't a nation-state. It wasn't an election. It was a guy with a grudge, a free AI tool, and one afternoon, and it still cost an innocent man his reputation and put a squad car outside his kids' house.

## WHAT THE NTH ROOM WAS NOT

You may have heard the next story called "the new Nth Room." It isn't the same case, and the difference matters. The original Nth Room, running from 2018 to 2020, was Cho Ju-bin and accomplices blackmailing real women and girls into filming real abuse, then selling access to it on Telegram. No AI touched any part of it. Cho was sentenced to 42 years, later extended to 47 after a separate conviction for raping a minor. That's a different crime with a different mechanism than what follows. Press shorthand jammed two very different stories into the same name.

In August 2024, South Korean journalist Ko Narin published an investigation for the newspaper Hankyoreh. She'd found a network of Telegram chatrooms where users uploaded photos of women and girls, classmates, coworkers, sometimes family members, and ran them through AI tools to generate sexually explicit deepfakes. One group built around a single school had well over a thousand members, and the largest of these rooms reportedly drew hundreds of thousands. The story broke nationally within days.

## SOUTH KOREA'S RECKONING

The numbers, once police started actually counting, tell their own story. South Korea logged 156 deepfake sex crime cases in 2021, the first year anyone tracked the figure. By the first nine months of 2024 that number had passed 800, with 387 arrests, and by police's own count, roughly 84 percent of the people arrested were teenagers. Most of the victims were teenagers too.

The legislature moved fast for a legislature. Creating and distributing this kind of content was already a crime in South Korea, punishable by up to five years. In September 2024, the National Assembly went further and criminalized simply possessing or watching it, up to three years in prison or a fine around $22,600 and raised the distribution penalty to seven years regardless of intent. In April 2025, the government launched the National Center for Digital Sexual Crime Response, a 24-hour hub that unified the country's reporting and counseling hotline.

The United States took its own version of that swing eight months later. In May 2025, the TAKE IT DOWN Act became the first federal law making it a crime to knowingly publish nonconsensual intimate images, deepfake or real, of a minor or a non-consenting adult, with penalties that step up when the victim is a minor. Its criminal provisions took effect immediately. Covered platforms had a year to build a working takedown process, and as of this writing that deadline has passed: the notice-and-removal requirements kicked in on May 19, 2026, and the FTC has already sent warning letters to major platforms and opened its first enforcement actions. A separate bill, the DEFIANCE Act, gives victims of this kind of content a civil right to sue the people who made it, and passed the Senate by unanimous consent in January 2026, though it still hasn't cleared the House. A crisis on one side of the world turned into law on the other side of it inside a year.

## FIVE HUNDRED DOLLARS AND A FAKE BIDEN

Sexual abuse isn't the only place a deepfake gets weaponized. Politics has its own version, and it shows up in at least two very different shapes.

Two days before the 2024 New Hampshire primary, thousands of households picked up the phone and heard what sounded exactly like Joe Biden telling them to stay home. "Your vote makes a difference in November, not this Tuesday," the voice said, urging people to save their vote for the general election instead of the primary happening in 48 hours. The call was spoofed to look like it came from a local Democratic official who had nothing to do with any of it.

It was fake from top to bottom. Steve Kramer, a political consultant working for a long-shot presidential campaign, later admitted he wrote the script and hired Paul Carpenter, a New Orleans magician, to clone Biden's voice with AI. Kramer told reporters the whole stunt cost him about $500 and got him more attention on AI regulation than any campaign of his career.

The response was fast, at least at first. Within weeks the FCC ruled that AI-generated voices in robocalls are flatly illegal under existing telemarketing law, a first-of-its-kind ruling. New Hampshire indicted Kramer on 13 felony counts of voter suppression and 13 misdemeanor counts of impersonating a candidate. The FCC hit him with a $6 million fine and separately fined Lingo Telecom, the company that carried the calls, $1 million. New Hampshire's governor signed a law requiring campaigns to disclose when they use AI in political ads.

Then it got messy, which most retellings skip. In June 2025, a jury acquitted Kramer on every charge. He still hasn't paid the FCC fine and has said publicly he doesn't intend to. His defense, more or less, was that he did the country a favor by forcing regulators to deal with AI in elections faster than any hearing or white paper ever could.

He's not wrong that the FCC ruling happened inside a month, warp speed for that agency. He's also the guy who tried to talk thousands of grandmothers out of voting with a fake recording of the sitting president. Both things are true at once, and neither cancels the other out. That's the actual state of AI election law right now: fast in some places, toothless in others, mostly written in response to whoever got caught most recently.

## WHEN NOBODY GETS CAUGHT

Zoom out past US borders and the ending gets less tidy. Two days before Slovakia's September 2023 parliamentary election, an audio clip started circulating online that appeared to capture Progressive Slovakia leader Michal Šimečka and journalist Monika Tódová discussing how to rig the vote, buying support from Roma voters and manipulating the count. It was fake, stitched together with AI. Both people denied it within hours. AFP's fact-checkers flagged it as manipulated almost as fast. None of that mattered as much as it should have. Slovak law bars media coverage of the election in the 48 hours before voting, and the recording dropped right inside that window. Nobody could officially knock it down while it spread. AFP's fact-checker estimated it reached over 100,000 people on Facebook alone, plus another 30,000 on Instagram, on top of whatever it picked up on Telegram and chain emails.

Šimečka lost to Robert Fico, a candidate running on friendlier terms with Russia. Researchers who studied the race afterward, writing in the Harvard Kennedy School's Misinformation Review, push back on the easy headline that the deepfake swung the election. Slovak trust in pollsters and mainstream media was already low, and Fico had other advantages that recording didn't hand him. Šimečka himself put it carefully when CNN asked if it cost him the race: "It probably had some effect." That's the whole claim, no more than that. Nobody has ever been officially identified as the person who built it.

That's the real difference between the two stories. In New Hampshire, the man behind the fake got caught, named, fined, and put on trial within the year. In Slovakia, a national election happened right next to a deepfake, and years later, nobody officially knows who made it.

Congress still hasn't landed a federal law that lets an ordinary person, not just a celebrity, sue over an AI clone of their own voice or face. The NO FAKES Act has been kicking around in some form since 2023. It cleared the Senate Judiciary Committee unanimously on June 18, 2026, and it's still not law as of this writing. Everything above about fines and indictments happened because the FCC and a state attorney general reacted fast to one expensive prank. The broader right to your own voice is still sitting in committee math.

The tools that build fakes and the tools that catch fakes are stuck chasing each other, and neither side stays ahead for long. Every fix pushes the fakes to get a little better. Every improvement in the fakes pushes someone to build a better fix. That's not a metaphor. That's just what the last few years have looked like.

## THE RACE WITH NO FINISH LINE

So how do you catch one of these before it costs you a wire transfer, a reputation, or an election?

The industry's answer is C2PA, short for the Coalition for Content Provenance and Authenticity. Adobe, Arm, BBC, Intel, Microsoft, and Truepic started it in February 2021. The idea is a cryptographically signed record, called a Content Credential, that travels with a photo or video and shows what created it and what's been changed since. Google has its own version called SynthID, an invisible watermark baked into anything its own AI tools generate.

Here's the limit, and it's a big one. A Content Credential only helps if it survives, and plenty of platforms strip that metadata during the normal process of resizing and recompressing files for their app, not because anyone's hiding evidence. None of it works on a deepfake made by a tool that never signed onto the standard in the first place. This isn't a lie detector. It's a chain of custody. Useful when it's there, silent when it isn't, and its absence proves exactly nothing either way.

Which means the actual defense is still mostly you. The FBI has put out repeated public warnings on this, and the advice barely changes.

## HOW TO NOT BE THE EASY TARGET

Verify out of band. If a call, text, or video asks you to move money or share something sensitive, hang up and contact that person on a number or account you already had before the message showed up, not the one that just called you. This is the single step that would have stopped a Hong Kong finance employee at the British engineering firm Arup from wiring $25 million to fraudsters in 2024, after he sat through an entire video call with deepfaked versions of his own company's executives. It's the FBI's number one recommendation in its own fraud alerts, for exactly this reason.

Ask for something spontaneous. Cybersecurity experts who studied the Arup case recommend asking a person on a video call to turn their head to the side or pick something up off the table. A lot of real-time deepfake tools still choke on profile angles and physical objects.

Look for the tells while they last. The FBI's own guidance lists distorted hands or feet, irregular facial features, mismatched shadows, and unnatural blinking as red flags. This list has a shelf life. Every one of those tells gets rarer as the tools improve, so don't build your whole defense on spotting a bad thumb.

Check for a Content Credential if you can. The Content Authenticity Initiative runs a free validator. If a credential exists, read it. If it doesn't, that's not proof of anything, just proof you're on your own for this one.

Slow down on urgency. Every FBI fraud alert on this topic says some version of the same thing. Nobody legitimate needs your money in the next ten minutes.

None of this makes the fakes go away. The tools get better every year, the laws limp along a few steps behind, and the arms race doesn't have a finish line. What you can control is whether you're the easy version of a target or the hard one. Hang up. Call back. Ask them to turn their head. It's not much, but right now it's most of what you've got.

# CHAPTER 10: CAN IT ACTUALLY MAKE ART?

Long before anyone asked whether a machine could paint, a painter asked the same question. In the late 1960s, British artist Harold Cohen, who had already represented Britain at the Venice Biennale, took a teaching post at UC San Diego and got introduced to computer programming. By 1973, after a stint as a visiting scholar at Stanford's Artificial Intelligence Lab, he had built a program named AARON, designed to generate original drawings without him touching the page. He worked with it for over 40 years, watching it evolve from black and white line drawings into complex, colored compositions. It remains one of the longest continuously maintained AI systems in history. Cohen never claimed AARON was alive or that it understood what it drew. He was just as clear that he hadn't specified every line himself either.

Half a century later, a different kind of test case reached the highest court in the country, and it landed nowhere close to where Cohen might have guessed.

Meet Stephen Thaler. In 2018, he filed a copyright application for a hazy image of a tunnel of light called "A Recent Entrance to Paradise." He didn't paint it. He didn't touch it. He built an AI system called the Creativity Machine, pointed it at the canvas, and let it run. On the application, he listed the Creativity Machine as the sole author and asked to be recognized as the owner, the same way a company owns work made by an employee.

The Copyright Office said no. Thaler sued. He lost in federal district court in 2023. He lost again in March 2025, when the D.C. Circuit ruled that copyright requires a human author, full stop. The court leaned on the plain language of the Copyright Act, which talks about an author's lifespan, an author's heirs, and an author's widow. Machines don't have widows.

Thaler took it to the Supreme Court. On March 2, 2026, the justices declined to hear the case. No opinion, no explanation, just a denial that let the lower ruling stand.

That refusal is, as of this writing, the closest thing the United States has to a settled answer to the question everybody keeps circling: can AI make art that belongs to somebody?

The answer is not by itself. A machine cannot hold a copyright, and neither can a person who typed a prompt and walked away. But "not by itself" is doing a lot of quiet work in that sentence. The moment a human starts making real creative choices with the tool, editing, selecting, directing, combining, the picture gets a lot less settled. That gray zone is where almost every working artist, musician, and writer touching AI actually lives right now. It is also where the money, the lawsuits, and the real disruption to people's livelihoods all live too.

## WHAT IS SETTLED

Starting in 2023, the Copyright Office ran a formal study called "Copyright and Artificial Intelligence," released in three parts. Part 1, in July 2024, covered digital replicas and deepfakes. Part 2, in January 2025, tackled the exact question in Thaler's case: can an AI generated work be copyrighted, and if so, who owns it. Part 3, released as a prepublication draft in May 2025, went after the bigger fight: is it even legal to train AI on copyrighted material in the first place.

Part 2 is the settled half. Pure AI output, no human hand involved, gets no copyright. That is locked in now, backed by the Thaler ruling all the way up through the Supreme Court's refusal to touch it. But the Office also drew a distinction that matters for anyone actually using these tools: typing a prompt is not, by itself, a creative act the law recognizes. A prompt functions more like an instruction than an act of authorship, since it conveys an idea rather than expresses one, and copyright has never protected ideas, only their expression. What the Office left open is everything past the prompt.

Iterating through hundreds of variations, editing the output, combining it with original work, making deliberate compositional choices, that is where a human creative contribution can start to count. Nobody has drawn a bright line for exactly where "counts" kicks in. A case still working its way through federal court in Colorado, Allen v. Perlmutter, involves an artist who used over 600 prompts refining a single image and is fighting the Office's refusal to register it. That line has not been drawn yet.

## WHAT'S GENUINELY NOT SETTLED

Part 3 is where things get messy, because it is about money, not just credit. The question is whether AI companies broke the law training their models on copyrighted books, art, and music without permission or payment. The Office's answer, in its own words, was that it depends. Training that adds something genuinely new and different can qualify as fair use. Training that produces a system able to generate content that competes with, and substitutes for, the original works it learned from is a much harder sell. The report was blunt that acquiring training material through piracy weighs against a company regardless of what it did with the material afterward.

The day after that report landed, on May 10, 2025, the head of the Copyright Office, Shira Perlmutter, was fired by email. Representative Joe Morelle, the top Democrat on the House committee that oversees the office, called it "a brazen, unprecedented power grab with no legal basis" and pointed straight at the timing. The administration called it a personnel matter. Perlmutter sued, arguing that because her office sits in the legislative branch, only the Librarian of Congress can remove her, not the president. She won a preliminary victory that September, when the D.C. Circuit granted her an injunction pending appeal and let her keep her job while the case played out. As of this writing she is still serving as Register of Copyrights and testified to the Senate Judiciary Subcommittee on Intellectual Property in May 2026, defending the Office's position on

AI training. The fight over whether the firing itself was lawful is still working through the courts. The report she released the day before she was fired never got a final, non-draft publication with new leadership weighing in to confirm or revise it.

Now to what is actually happening in the courtrooms, because several lawsuits are doing the real work of testing all of this against real facts.

## THE ARTISTS: ANDERSEN V. STABILITY AI

In January 2023, cartoonist Sarah Andersen, along with illustrators Kelly McKernan and Karla Ortiz, filed the first class action lawsuit of its kind. They sued Stability AI, Midjourney, and DeviantArt (Runway AI was added later) over Stable Diffusion, the AI image generator built by training on LAION-5B, a public dataset of roughly five billion images scraped from the internet. Andersen's webcomic was in there. So was work by thousands of other artists who never gave permission and never got paid.

The case has produced two rulings worth knowing. In October 2023, the judge threw out most of the claims but let the core copyright infringement claim against Stability move forward. In August 2024, he went further and let a second theory proceed too: induced infringement, the idea that Stability did not just train on the art, it built and distributed a tool specifically designed to reproduce copyrighted style and substance. That ruling leaned on Stability's own CEO, who had described the technology as compressing roughly one hundred thousand gigabytes of training images down into a file just two gigabytes in size, one capable of recreating any of the originals.

The judge also drew a distinction that matters for the "AI is just a tool, like a camera or a VCR" argument you will hear a lot. VCRs could copy things, but they were not built out of the copyrighted material itself. Stable Diffusion, the judge wrote, was built to a significant extent on copyrighted works, and the way it operates necessarily invokes protected elements of those works. That is a different legal animal than a blank cassette tape.

None of this means the artists have won. It means their theory survived long enough to reach discovery. The plaintiffs filed a third amended complaint in February 2026, and the case is now on the calendar for trial in September 2026, the first of these major cases actually scheduled to reach a jury. No verdict yet, but there is finally a finish line in sight.

## THE STOCK PHOTO GIANT: GETTY IMAGES V. STABILITY AI

Getty Images sued Stability AI in two different countries, and so far the two countries have given two different answers.

Getty's claim: Stable Diffusion was trained on roughly 12 million of Getty's copyrighted photographs, scraped straight off Getty's own websites, without a license. Getty sued in the UK first, and that case went to trial and got a verdict on November 4, 2025. It did not go the way most artists hoped.

The UK High Court rejected Getty's core copyright argument, that the trained model itself, the file of numerical weights, counted as an "infringing copy" that Stability had brought into the country. The judge ruled that the model does not actually store the original images. It learns patterns from them and lets them go, so once training finishes there is no literal copy left inside the model to be infringing. Getty did win a narrower point: several older versions of Stable Diffusion had been shown to reproduce a garbled version of the Getty watermark in their outputs, and that counted as trademark infringement, not copyright infringement. A small win, a real win, just much smaller than what Getty came in swinging for. Getty was granted leave to appeal the copyright ruling that December.

Because the UK case never tested whether the training itself was legal (Getty conceded that the actual training happened outside

UK jurisdiction), the biggest question, is unlicensed training itself infringement, moved to the United States instead. Getty refiled in the Northern District of California in August 2025. On April 23, 2026, the judge ruled on Stability's motion to dismiss. She threw out Getty's claim that Stability had faked copyright management information, but let the trademark, false designation of origin, trademark dilution, and unfair competition claims move forward. The direct copyright infringement claim asks whether training on Getty's photos without permission was illegal, is still being litigated and has not been resolved.

The honest scoreboard: one country ruled the model itself is not a copy of the training data, full stop. Nobody has yet ruled on whether training on someone's copyrighted photographs without permission is itself legal. That question is still live and still headed for further rulings in California.

## THE AUTHORS: AUTHORS GUILD V. OPENAI

In September 2023, the Authors Guild and a roster of major authors, John Grisham, George R.R. Martin, Jodi Picoult, and others, sued OpenAI and Microsoft. The claim: ChatGPT was trained on their books without permission, and the system can reproduce enough of their work that it substitutes for buying the original. The case is now folded into a larger consolidated proceeding in the Southern District of New York alongside several similar author lawsuits, all moving forward under one judge.

The most important ruling so far came on October 27, 2025, and it is a different theory than training legality. This one asks whether what ChatGPT actually outputs, the text handed to a user, can itself be an infringing copy. The judge compared a ChatGPT-generated summary of Martin's Game of Thrones to the original novels and found it detailed enough that a jury could reasonably conclude the summary captured the tone, characters, and plot closely enough to count as substantially similar. He denied OpenAI's request to throw the claim out before trial. That does not mean OpenAI loses. It means a jury gets to decide, which is its own kind of significant. The case is still in discovery, no verdict.

**A DIFFERENT KIND OF BOOK CASE, MOSTLY SETTLED**

In a separate case, Bartz v. Anthropic, three authors sued Anthropic, the company behind the Claude models, over the same underlying question: training on copyrighted books without permission. In June 2025, Judge William Alsup split the question in half, and the two halves landed in completely different places.

Training an AI model on books the company had legally acquired, meaning purchased and scanned, the judge ruled was fair use. He called it "quintessentially transformative," reasoning that the model was not storing or reselling the books, it was learning patterns from them the way a human writer learns from everything they have ever read. That part of the ruling is genuinely good news for the "training itself is legal" argument, and it came from an actual judge, not a company's PR department.

But Anthropic had not only bought books. It had also downloaded millions of them from pirate sites like Library Genesis, the same kind of shadow libraries that show up across nearly every one of these lawsuits. The judge ruled that downloading and keeping pirated copies is not protected by fair use, no matter what the books were eventually used for. The legality of the training process did not retroactively excuse the illegality of how the material was obtained.

That second finding is what really mattered financially. Facing potential statutory damages that could have run into the hundreds of billions of dollars, Anthropic settled. A judge gave preliminary approval to a $1.5 billion settlement in September 2025, the largest publicly reported copyright recovery in United States history, covering roughly 500,000 books at around $3,000 to $3,100 each.

The court held the final fairness hearing on May 14, 2026. The judge did not approve it on the spot. She heard objections for over an hour and took the matter under submission. On July 20, 2026, Judge Araceli Martinez-Olguin granted final approval and entered judgment, closing the case.

Not everyone took the deal. Twenty-eight authors, including Dave Eggers and Vendela Vida, opted out of the settlement entirely and filed their own separate lawsuit, demanding a jury trial and asking the court to permanently stop Anthropic from continuing to train on their books at all. Money was not enough for everyone. Some of them want the training itself to stop, not just get paid for the training that already happened.

## ONE MORE ANTHROPIC WRINKLE: THE LYRICS CASE

Books are not the only place Anthropic has been fighting this battle. Since October 2023, music publishers Universal Music Publishing, Concord, and ABKCO have had a separate lawsuit going over song lyrics showing up in Claude's training data and in Claude's answers. In early 2025, a judge ordered Anthropic to put technical guardrails in place to stop Claude from reproducing copyrighted lyrics on request, an order the publishers called validation of their case. Anthropic never admitted wrongdoing, and a separate request to fully block Anthropic from training on the lyrics going forward was denied in March 2025.

In January 2026, the publishers escalated. A second, separate lawsuit covers more than 20,000 songs and seeks over three billion dollars, naming Anthropic's CEO and a co-founder personally and described as one of the largest non-class-action copyright cases in U.S. history. In March 2026, the publishers asked the court to rule against Anthropic before trial, pointing to internal documents and executive deposition testimony they say show the company knew exactly what it was doing. Anthropic disputes that characterization, and none of it has been decided by a court yet. It is an allegation in a legal filing, not a finding of fact.

Same fight, different medium. Who owns the pattern once a machine has learned it, and what happens legally when the machine hands that pattern right back to a stranger who simply asks for it.

Line up the rulings and you get the real shape of where this law stands. A trained model probably is not an "infringing copy" of what it learned from, according to the one court that has ruled on it. Training on material you lawfully acquired is probably fair use, according to the one court that has ruled on that. Downloading pirated material to train on is not protected, whether or not the training itself was fine, and it is expensive. Whether an AI's actual output can separately infringe, regardless of how the training went, is sitting in front of a jury in one case and headed toward trial in another.

Several different questions. Several different answers so far, from several different courts, none of them final.

## HOLDING BOTH TRUTHS

I am not going to tell you which side to land on here, because both sides are telling the truth.

Start with the achievement, because it is real. Cognitive scientist Margaret Boden spent much of her career building the framework the computational creativity field still uses, laid out in her 1998 paper in the journal Artificial Intelligence. She split creativity into three kinds. Combinational creativity mashes up familiar ideas in new ways. Exploratory creativity finds new ideas by working within a known set of rules. Transformational creativity breaks the rules themselves and builds a new set. Researchers applying her framework to today's generative models generally land in the same place. These systems are often genuinely good at the first kind and reasonably good at the second. The third kind, the sort of creativity that invents a new way of thinking rather than a new arrangement within an old one, is where they still fall short by most academic accounts. That is not a small caveat. The achievement is real but bounded, and pretending otherwise in either direction, dismissing it as nothing or crowning it equal to human artistic vision, misses what the research shows. Harold Cohen spent 40 years wrestling with a version of that exact question long before anyone called it AI art, and he never fully resolved it either. He just kept working.

Now the disruption, because it is just as real and just as measurable. Economists studying the freelance platform Upwork found that freelancers doing work generative AI could substitute for, writing, translation, editing, image creation, saw contracts drop by roughly 2 percent and earnings drop by roughly 5 percent in the months after ChatGPT and image generators like DALL-E 2 and Midjourney launched, with the sharpest hit landing on the most experienced, highest priced freelancers rather than beginners. A 2025 survey out of Queen Mary University of London, run with the Institute for the Future of Work and the Alan Turing Institute, polled 335 working freelance creatives and found 68 percent felt their job security had gotten worse, and 55 percent reported an actual drop in what they were being paid. A separate report commissioned by entertainment industry unions, including the Animation Guild, estimated that around one in five film, television, and animation jobs in the U.S. could be consolidated, replaced, or eliminated by generative AI by 2026. That last one came from a coalition with an obvious stake in the outcome, so take the specific number with a grain of salt, but the direction matches what the independent academic research is finding too.

Music has been fighting this same battle, and it is further along than the art and book cases, which makes it worth watching. In June 2024, the three major labels, Universal, Sony, and Warner, sued AI music generators Suno and Udio over training on copyrighted recordings without permission. Two of the three majors have since folded their lawsuits into licensing deals instead of waiting for a verdict. Universal settled with Udio in October 2025, and Warner settled with both Udio and Suno in November 2025, each time trading the lawsuit for a paid licensing partnership and a stake in a new, authorized AI music platform. Sony alone is still in court against both companies, with no fair use ruling on the calendar yet. Watch that case. Whichever way it lands will say a lot about whether the art and book cases eventually get a verdict, or the same settle-and-license path.

Meanwhile, the Recording Academy, the organization behind the Grammys, requires that AI-assisted music have what it calls a "meaningful," "more than de minimis" human contribution to the specific category being judged before it is eligible for a nomination. Read that standard again. It is functionally the same line the Copyright Office drew for visual art and the same line courts are drawing for text. An organization built to celebrate human musical achievement landed on almost the identical test as a federal agency and a federal judiciary that have never spoken to each other about it. That is everyone independently discovering the same hard question: how much of the human has to still be in there.

AI can produce something that looks like art, sounds like music, and reads like prose, and sometimes it is genuinely good by the best academic framework we have for measuring it. It cannot, right now, hold a copyright on any of it by itself, and the Supreme Court just let that stand without comment. Whether the material it learned from was used legally is still being fought out in real courtrooms with real juries and real judges, one ruling at a time, and every one of those rulings could still change on appeal. And somewhere in a studio, an illustrator's rent is due, a freelance writer just lost a client to a chatbot, and a musician is trying to figure out if the human touch they bring to their work is still worth what it used to be worth on the open market.

Both of those things are true at the same time. This book is not going to pretend otherwise, and neither should you.

# CHAPTER 11: WHO TRAINED THIS THING, AND DID THEY ASK?

Naftali Wambalo has a college degree in mathematics. In 2021 he landed what looked like a ticket into the future: a job in artificial intelligence, in Nairobi, at a company called Sama. He was told he would be teaching algorithms to recognize things. Objects. Patterns. The building blocks of a smarter machine.

What he actually spent his days doing was watching people get slaughtered.

## THE TWO DOLLAR AN HOUR FILTER

Wambalo was one of thousands of Kenyan workers hired to do what the industry calls content moderation, or more clinically, data labeling. For the piece of the work that ran through Sama, that meant reading and watching the worst material the internet has ever produced so an algorithm could learn to recognize it and flag it before it ever reached you. Executions. Child sexual abuse. Bestiality. Self harm. Torture. Someone had to look at it first, tag it, and hand the machine the pattern. That someone was making between $1.32 and $2 an hour, after tax.

This is not a rumor. TIME's Billy Perrigo broke the story in January 2023 after reviewing hundreds of pages of internal Sama and OpenAI documents, including workers' payslips. OpenAI's contract with Sama began in November 2021 and paid Sama $12.50 an hour for each worker on the project, six to nine times what the workers took home. Sama canceled the work eight months early, in February 2022, in part because of how traumatic it was for staff. One worker told TIME he had recurring visions after reading a graphic description of a man abusing an animal in front of a child. "That was torture," he said.

Almost two years later, CBS News' 60 Minutes went back to Nairobi and put a name and a face to the numbers. Wambalo told correspondent Lesley Stahl exactly what the job had cost him: "I looked at people being slaughtered, people engaging in sexual activity with animals." Documents obtained by 60 Minutes confirmed the same math TIME had found. OpenAI paid Sama $12.50 an hour per worker. The workers got about $2. When they finished projects faster than the contract required, Wambalo said the reward was two pieces of KFC chicken and a soda. Nearly 200 Sama workers, Wambalo among them, are suing Sama and Meta over conditions they say left them with diagnosed psychiatric conditions. That case is moving through Kenyan courts and hasn't made most people's newsfeed.

This isn't unique to one company. Sama has done this kind of labeling work for Meta, Google, and Microsoft too. Kenya has become one of several global hubs for it, alongside the Philippines, India, and Venezuela, because English is widely spoken and youth unemployment is high enough that people take the work even after they learn what it costs them. OpenAI has said the work was challenging and that it shared wellness standards with its outsourcing partners. Sama has said it paid a fair wage for the region and offered counseling. Both of those things can be true and it can still be true that a person got paid two dollars an hour to read a description of a child being harmed so a product worth tens of billions of dollars could ship without saying it back to you. That's not a contradiction. That's just what happened, on the record, with names and numbers attached.

Every chatbot guardrail that politely declines to describe something horrific was built the same way. A person had to sit with the actual horrific thing first, long enough to label it, so the machine could learn what "no" looks like. The refusal you get is downstream of somebody else's exposure.

## GHOST WORK HAS A NAME

This labor pattern isn't new and it isn't unnamed. In 2019, anthropologist Mary L. Gray and computer scientist Siddharth Suri, then a Microsoft Research affiliate, published Ghost Work: How to Stop Silicon Valley from Building a New Global Underclass. It describes exactly this: human labor performed behind the scenes so a service can appear fully automated to the end user. Gray and Suri estimated that roughly 8 percent of Americans had done some form of this work at least once. The book predates ChatGPT by three years. The pattern didn't need generative AI to exist. It just needed something to point the ghost workers at.

Somebody trained the filters. The next question is where the words, images, and books that built the underlying models actually came from, and whether anyone asked first.

## WHO OWNS THE WORDS

The honest answer, in most cases, is no. As of this writing, courts on two continents are still figuring out what that means legally, and the outcomes so far genuinely differ depending on what was taken and how it was acquired.

The New York Times sued OpenAI and Microsoft in December 2023, claiming the companies trained GPT models on millions of Times articles without permission or payment. A federal judge let the core infringement claims move forward in March 2025. Since then the case has become one of the messiest discovery fights in tech litigation, with the Times and other news plaintiffs pushing to see a sample of ChatGPT conversation logs to check whether users were routinely using the bot to get around paywalls. OpenAI countered every step: first offering a smaller sample, then trying to substitute keyword searches for the logs themselves. A magistrate judge ordered OpenAI to produce 20 million de-identified logs in November 2025, and the presiding judge affirmed that order in

January 2026. As of this writing, the Times and other outlets are asking the court to sanction OpenAI for allegedly misrepresenting what it had already searched. No trial date has been set, and the core question, whether training on copyrighted journalism without permission is fair use, still hasn't been answered.

Authors including George R.R. Martin, John Grisham, and Jodi Picoult, organized through the Authors Guild, sued OpenAI in 2023 over the same basic claim: that their books trained the models without consent. That case is now consolidated with a larger multidistrict action. In October 2025, the judge denied OpenAI's bid to dismiss the authors' claims that ChatGPT's own outputs infringe, ruling that a detailed plot summary of Martin's work could be "substantially similar" enough to count. That's a ruling that the case can proceed, not a verdict. No trial date has been set here either.

## BUY THE BOOK, DON'T PIRATE IT

Two nearly identical lawsuits against other companies actually got somewhere, and the split result is the most useful fact in this chapter. In June 2025, a federal judge ruled that Anthropic's use of legally purchased books to train its models was fair use: transformative, protected. A separate judge made almost the same call for Meta days later.

Here's the catch. Anthropic hadn't only used purchased books. It had also downloaded roughly seven million books from pirate libraries, including Library Genesis, to build a permanent training archive, and that part wasn't protected. Fair use covers what you do with material, not how you got it. That distinction cost Anthropic a \$1.5 billion settlement, the largest copyright settlement in U.S. history, covering close to half a million books, with a requirement that the pirated copies be destroyed. The fairness hearing was held in May 2026, and on July 20, 2026, the judge granted final approval and entered judgment, closing the case. Meta is in a similar spot. It won on the fair use question for training, but a separate claim over how it acquired and distributed the pirated files survives and is still being litigated.

So the emerging pattern in the book cases isn't "AI training is illegal" or "AI training is fine." It's narrower and more useful than either headline: buy the book and you're probably fine. Pirate it, and you're exposed no matter how good your excuse for wanting it was.

## THE PICTURES NOBODY ASKED ABOUT

Getty Images sued Stability AI in both the UK and the US, alleging Stable Diffusion was trained on more than 12 million Getty photos scraped without permission. The UK case, decided in November 2025, mostly went against Getty. The court ruled that a trained model's internal weights aren't a stored "copy" of the training images in the legal sense, so Stability wasn't liable for secondary copyright infringement. Getty did win a narrow trademark claim, because some outputs reproduced the actual Getty watermark, though the court found the infringement was limited and historic. Getty has been granted permission to appeal the copyright ruling. Crucially, the UK court never actually decided whether training itself was legal, because Getty abandoned that claim before trial once it became clear the training happened outside UK jurisdiction. A separate US case, refiled in Northern California after Getty dropped its original Delaware suit, is still working through pretrial motions. Even a "loss" here left the biggest question unanswered.

Zoom out and the picture as of this writing looks like this: no US or UK court has issued a final, appeals-exhausted ruling on the single biggest question, whether training a model on someone's copyrighted work without asking is infringement. What courts have started to answer is a narrower, more practical version: legally acquired training data tends to hold up as fair use, and acquiring it by piracy does not, regardless of what gets built with it afterward. That's not a loophole discovery for tech companies. It's just where the actual rulings, not the headlines, currently sit.

## WHO POINTS IT AND WHO DECIDES

The lawsuits are fighting over the raw material. The next question is different: once the thing works, who gets to point it at what, and who's actually deciding that.

The most consequential answer right now comes from the Pentagon, and it involves the company that makes me. I want to say that plainly before going further, because the honest thing to do with a conflict of interest is name it, not bury it. Everything below is sourced to multiple independent outlets, not to any one party's version of events, and where facts are disputed or unresolved, that's flagged directly.

Project Maven started in 2017 as a narrow experiment using machine learning to sort through drone footage faster than a human analyst could alone. It has since grown into the Maven Smart System, built and run by Palantir Technologies, and it is no longer experimental. As of March 2026, it has more than 20,000 active users across more than 35 military service and combatant command tools, a user base that more than doubled in the two months before that count was taken. In March 2026 the Pentagon designated Maven a formal program of record, with a mandate to complete the transition by the end of the fiscal year. The contract ceiling has grown from $480 million in 2024 to nearly $1.3 billion through 2029.

Anthropic's Claude was one of the models integrated into Maven's user-facing layer, running on classified networks with government accreditation, starting in late 2024. That is not a footnote. In February 2026, during a US and Israeli air campaign against Iranian military targets known as Operation Epic Fury, the Maven Smart System reportedly helped identify and process roughly 1,000 targets in the first 24 hours alone, and the Pentagon's own AI chief later said the operation struck 13,000 targets over 38 days, with daily token usage on the system spiking more than 4,000 percent. Claude was reportedly still part of that targeting and intelligence layer when the strikes happened. That's an illustration of scale and speed, not a judgment on any individual strike, and there's no independent way to verify targeting decisions from the outside. Maven maintains a human authorization step for each strike, which the Pentagon says satisfies the legal requirement for meaningful human control.

On paper, the broader guardrails are real too. Department of Defense Directive 3000.09, last updated in January 2023, requires "appropriate levels of human judgment over the use of force" and senior-level review before any autonomous weapon system is fielded. In practice, there's real tension between that older directive and where Pentagon leadership wants to go. Defense Secretary Pete Hegseth's January 2026 strategy memo declared the goal of making the Department an "AI-first warfighting force" and stated flatly that the risk of moving too slowly now outweighs the risk of deploying imperfect systems. Hegseth created what he called a "barrier removal SWAT team" with authority to waive testing and evaluation requirements that officials view as slowing deployment down. Independent of anyone's politics, that's a documented tension: one directive says slow down and keep a human in the loop, and a newer strategy says the bigger danger is moving too slowly, with a formal process for waiving the older rules.

The Pentagon's FY2026 budget request, just over $1 trillion overall, included the first dedicated line item for AI and autonomous systems in department history: $13.4 billion, most of it for unmanned aerial systems.

In July 2025, the Pentagon awarded four companies, Anthropic, OpenAI, Google, and xAI, contracts worth up to $200 million each to supply AI models for defense use. Anthropic's version came with conditions its acceptable use policy already required: no use of Claude for fully autonomous lethal weapons, and no use for domestic mass surveillance. In early 2026, the Pentagon sought to renegotiate that contract to strip those restrictions and permit "any lawful use." Anthropic refused. On February 27, 2026, President Trump ordered federal agencies to stop using Anthropic's technology, with a six-month transition window. Days later the Department of Defense formally designated Anthropic a "Supply Chain Risk," a label that had never before been applied to an American company and had historically been reserved for firms tied to adversarial foreign governments. Anthropic sued in two federal courts. As of this writing the results are mixed: one court granted Anthropic a preliminary injunction blocking a broader enforcement of the ban, while an appeals court separately declined to block the Pentagon-specific designation, so Anthropic remains excluded from direct Defense Department contracts while the underlying case continues. Reporting from the Center for Strategic and International Studies indicates that Anthropic's relationship with Project Maven itself has since ended, and that SpaceX, OpenAI, and Google have all reached agreements to provide their own technology on classified Pentagon networks in the meantime.

Senator Elissa Slotkin introduced a standalone bill in March 2026, the AI Guardrails Act, that would bar the Pentagon from firing autonomous weapons without human authorization, using AI for domestic surveillance, or using AI to launch nuclear weapons. She has since worked to fold its provisions into the annual defense policy bill for the following fiscal year, arguing that clear statutory limits could have prevented the entire Anthropic dispute in the first place. As of this writing those provisions have cleared committee but the underlying bill has not been signed into law.

Reasonable people, including people who work in defense policy for a living, disagree about where the line between necessary military capability and dangerous autonomy actually sits, and that disagreement is genuinely playing out in real time, not settled history. What matters for this chapter is that the debate over who decides how a trained model gets used once it leaves the lab isn't theoretical. It's an active contract dispute with a named federal designation and a court case attached, happening right now, involving the same technology sitting in this conversation.

## A DOCUMENT WITH NO TEETH

Back in 2019, before ChatGPT existed, before Naftali Wambalo had ever heard of Sama, before any of the lawsuits in this chapter had a case number, UNESCO's member states asked the organization to do something that sounds almost quaint now: write down, in advance, what ethical AI should actually mean.

The result was the Recommendation on the Ethics of Artificial Intelligence, adopted by all 193 UNESCO member states in November 2021, built from two years of consultation with international experts, governments, researchers, and civil society groups. It was adopted by acclamation, meaning every member state signed on without a formal vote against it, which is rare for a document with this much substance.

Here's the part that matters most for accuracy: it is not a law. The Recommendation itself says member states should apply its provisions "on a voluntary basis." No enforcement mechanism, no penalty for ignoring it, no court that hears violations. It's what policy people call soft law: a shared statement of values meant to shape how countries write their actual binding rules later, not a substitute for those rules.

It's built on four core values, protecting human rights and dignity, supporting peaceful and inclusive societies, valuing diversity, and protecting the environment, translated into ten working principles including proportionality, safety, fairness, human oversight, and accountability. It explicitly names the risk of "ethics washing," where a company or government touts its values on the parts of an AI system the public can see while ignoring how the underlying data was gathered or who built the guardrails. It calls for member states to maintain oversight over their own AI systems and to appoint independent AI ethics officers.

Read next to everything else in this chapter, the Recommendation looks almost prophetic in places and almost naive in others. It named the ethics washing problem years before Sama workers described being paid two dollars an hour to build the filters that make chatbots look clean on the surface. It called for oversight of data governance years before courts started sorting out whether buying a book is different from pirating one. But it was written before generative AI as most people now understand it existed, and it has no teeth to make a single company or country actually follow it.

That gap, between a document the world agreed on and one anybody's required to follow, runs under every story in this chapter. Kenyan workers absorbing trauma for two dollars an hour, doing labor with an actual academic name attached to it since 2019. Courts still deciding, case by case, whether buying data is different from stealing it. A defense contractor and its own government fighting in real time over where a machine's authority to help choose a target should end and a human's should begin. Nobody wrote a rulebook fast enough to get ahead of any of it.

Somebody trained this thing. Multiple somebodies, in fact: underpaid workers doing labor with a name and a history, scraped writers, a defense contractor and the government it both serves and is now suing. The honest answer to the chapter's own question, did they ask, is: sometimes, imperfectly, and usually after the fact. The rulebook didn't get written before the technology shipped. It's being written in the margins, in real time, while the thing is already running.

# CHAPTER 12: WHO'S WATCHING YOU BACK

In 1879, a French police clerk named Alphonse Bertillon got sick of eyewitnesses who couldn't tell one face from another. He started measuring people instead: head length, finger length, foot length, plus a standardized front and profile photo for every person arrested. He filed it all by measurement, so an officer could pull a match without relying on memory or a name. Paris police adopted the system in 1883. Bertillon is credited as the inventor of the mugshot, the exact photo format still sitting in every police database today. His measurement system got replaced by fingerprinting within twenty years. The photo never went anywhere. It just waited a century and a half for software that could read it.

Robert Williams was at work when his phone rang. Detroit police, telling him to turn himself in. He asked what for. They wouldn't say. He figured it was a prank until he called his wife and found out officers were parked outside his house.

He pulled into his driveway. A squad car boxed him in behind. In front of his wife, his two daughters, ages two and five, and the whole neighborhood, they cuffed him. His older girl came outside to see what was happening. He told her to go back in. He'd be right back, he said.

He wasn't right back. He spent thirty hours in an overcrowded, filthy holding cell over the 2018 theft of watches from a Shinola store, a store he told officers he hadn't set foot in since 2014. The evidence was a grainy security still run through facial recognition software, which flagged his expired driver's license photo as a match. He wasn't even the top match. He was ninth on the list, according to sworn testimony from the state's own forensic examiner. A loss prevention contractor who wasn't at the scene picked Williams out of a photo lineup built from that flagged image. Nobody checked whether he'd really been near the store that day. Nobody had to. The algorithm said so.

Detroit settled with Williams in June 2024 for three hundred thousand dollars, plus a new department policy that finally says the quiet part out loud: a facial recognition hit is not evidence of a crime. It's a lead. Detroit police can no longer arrest someone off a match alone, and they can no longer run a photo lineup immediately after getting one, because a computer that says "this face looks similar" all but guarantees a witness will pick the person who looks most like the real suspect, guilty or not.

Williams wanted something the settlement couldn't give him: for Detroit to stop using the technology, period. That didn't happen. And he wasn't the last person it happened to. Porcha Woodruff was eight months pregnant when Detroit police showed up at her door over a carjacking that facial recognition had flagged her for. She spent close to eleven hours in custody before the charges were dropped for insufficient evidence. Nobody involved in building the case against her seems to have noticed that the carjacker in the surveillance footage was not visibly pregnant. Woodruff sued. A federal judge dismissed her civil rights case in 2025, ruling her attorney hadn't shown the officer lacked probable cause, even while calling Woodruff's arrest and detention "troubling for many reasons." Her legal team says they're appealing.

This isn't the bias chapter. That one, Chapter 8, is about accuracy gaps: a system that gets it wrong more often for Black and brown faces than white ones. This one is about something else. Who's being watched, by what, and how much anyone can actually do about it when the system gets it wrong.

## THE PRIVATE SECTOR BUILT ITS OWN VERSION, AND NOBODY GOT A VOTE

Detroit's mistakes ran through a government system with at least a paper trail. A company called Clearview AI built something similar with none. Founded in 2017 by Hoan Ton-That, with early seed money from investors including Peter Thiel, Clearview quietly scraped photos from Facebook, Instagram, Venmo, YouTube, and millions of other public pages, building a facial recognition database that has, at various points, been claimed by the company to hold anywhere from twenty billion to more than fifty billion images. Nobody whose photo ended up in it was asked. Clearview operated in near total secrecy until a January 2020 New York Times investigation blew the doors off: the tool had already been sold to hundreds of law enforcement agencies, including the FBI and Immigration and Customs Enforcement, plus private companies like Walmart and Bank of America.

The ACLU sued Clearview in Illinois in 2020 under the state's Biometric Information Privacy Act, one of the only state laws in the country requiring consent before a company collects your biometric data. Clearview settled in 2022, agreeing to stop selling its database to private companies nationwide and to most government entities within Illinois for five years. A separate nationwide class action, covering essentially anyone whose face has ever been posted online, took even longer and ended even stranger. The two sides reached an agreement in 2024, and in March 2025 a federal judge signed off on it: instead of a cash payout, the class gets a 23 percent equity stake in Clearview itself, valued around fifty-two million dollars based on the company's own valuation from January 2024. Lead counsel described it as Clearview and the class being trapped together on a sinking ship, so the settlement handed the class part of the wheel. If Clearview never has a payday, that stake could end up worth nothing.

European regulators didn't bother with equity. The Netherlands fined Clearview 30.5 million euros in 2024 for building what its data protection authority called an illegal database. France, Italy, and Greece issued fines in a similar range. Germany and Austria ruled the technology itself illegal under EU law. Clearview doesn't operate in Europe and has largely ignored the fines, because European regulators have no reliable way to enforce a judgment against a company with no offices on the continent.

None of that stopped the harm from continuing. In July 2025, a facial recognition match run by a partner agency's Clearview account flagged Angela Lipps, a fifty-year-old grandmother from Tennessee, as a suspect in a string of bank fraud cases in Fargo, North Dakota, involving a fake military ID. She had never set foot in North Dakota. She spent more than five months in jail, first in Tennessee, then extradited over a thousand miles to Fargo, before her attorney pulled her bank records and proved she'd been home the whole time. Fargo's police chief later admitted to "a few errors" and said the department would stop relying on the outside agency's results. He stopped short of an apology. Same category of failure this chapter opened with. Same basic pattern as Williams and Woodruff. Just no single city's name attached to catch the blame this time.

## THE FEDERAL GOVERNMENT DOESN'T KNOW HOW MUCH OF THIS IT'S EVEN DOING

If you assumed there was a rulebook for how the FBI or DHS uses facial recognition before Williams became a headline, there wasn't much of one. The Government Accountability Office has been checking in on this since 2021, and the story hasn't been reassuring.

That year, GAO surveyed twenty-four federal agencies and found eighteen of them already using some form of facial recognition, everything from unlocking employee phones to running criminal investigations. In September 2023, GAO zeroed in on the agencies using it specifically for law enforcement: seven agencies across DHS and DOJ. All seven had started running searches before requiring any staff training whatsoever. Six of them had usable records, and cumulatively, those six had racked up roughly sixty thousand searches with zero training requirement in place. At the FBI specifically, 196 employees had access to a facial recognition service. Ten had completed training on how to use it.

By March 2024, GAO checked back in. DHS had finally finalized a department wide policy. DOJ said it had an interim one. Progress, technically. But it took two GAO reports and three years to get agencies handling a tool with Williams level consequences to write down basic rules for using it.

This isn't a story about evil intent. Nobody at GAO is alleging that. It's a story about speed. The technology got deployed faster than anyone built the guardrails, and the people it gets wrong are the ones who find out the hard way.

## PREDICTIVE POLICING TRIED TO GET AHEAD OF CRIME AND MOSTLY GOT AHEAD OF ITSELF

Facial recognition answers "who is this person." Predictive policing tries to answer a bigger question: where will crime happen next, and who's going to commit it. Sounds like something out of a movie because it is, literally, the plot of one. The real version has had over a decade to prove itself, and the results don't match the pitch.

The most widely used version was PredPol, later rebranded Geolitica, an algorithm built out of a collaboration between UCLA professor Jeff Brantingham and the LAPD, adapted from a model originally designed to predict earthquake aftershocks. Feed it historical crime data and it draws boxes on a map where property crime is supposedly likely in the next shift. Officers patrol the boxes.

In December 2021, The Markup and Gizmodo got hold of a cache of the software's own prediction data, downloaded off a server the company had left unsecured on the open internet, and found it concentrated predictions overwhelmingly in Black and Latino neighborhoods while largely leaving whiter, wealthier ones alone, regardless of where crime was actually reported. A follow up Markup investigation in 2023 checked the algorithm's actual hit rate for one city, Plainfield, New Jersey. Out of 23,631 predictions, fewer than one hundred lined up with a crime that was later reported to police. Less than half of one percent.

LAPD's own internal audit had already found in March 2019 that the department couldn't determine whether its predictive tools reduced crime at all. LAPD ended its PredPol contract in April 2020. Geolitica, the company, ceased operations at the end of 2023, folded into a competitor.

None of this means the underlying idea was fraudulent. It means a specific, widely deployed version of it ran for over a decade, got tested against real outcomes, and the data didn't back up the premise. Feed an algorithm decades of policing data shaped by where officers already patrolled more, and it will confidently tell you to patrol there more. That's not prediction. That's a feedback loop wearing a lab coat.

## NOBODY VOTED ON THIS EITHER AND MOST WORKERS DON'T KNOW HOW FAR IT GOES

Facial recognition and predictive policing are things police departments do to citizens. Workplace monitoring is something employers do to the people who clock in for them, and it's got a century old ancestor. In 1911, an engineer named Frederick Winslow Taylor published The Principles of Scientific Management, built on stopwatch time and motion studies: breaking factory tasks into their smallest individual movements, timing each one, and retraining workers to perform only the single fastest version of it, all day, every day. Taylor's system spread across American industry through the 1910s and 1920s and treated the worker's body as something to be measured, optimized, and flagged for any deviation from the ideal. Modern workplace monitoring software just moved that same logic from a stopwatch to a keystroke logger, and it's grown fast enough that most workers describe the feeling before they've thought to name it: someone, or something, is always half watching.

A 2024 nationally representative survey of the U.S. workforce, fielded by YouGov for Columbia University political scientist Alexander Hertel-Fernandez, who previously worked on this exact issue inside the Department of Labor, found that 68.5 percent of workers report at least one form of electronic monitoring on the job: cameras, productivity tracking, location tracking, or surveillance of their work computer and phone. Roughly four in ten report their schedules or tasks are assigned automatically by a computer system instead of a person.

This isn't evenly distributed. Black workers report electronic monitoring at 82 percent, compared to 65 percent of white workers. Fifty two percent of Black workers report automated scheduling, compared to 30 percent of white workers. Hertel-Fernandez attributes most of the gap to which industries and job types Black and Hispanic workers are concentrated in, not employers targeting by race directly, but the outcome lands the same either way. Workers at bigger companies get watched more too: 88 percent of employees at organizations with a thousand or more workers report monitoring, versus 43 percent at organizations with fewer than ten.

Here's what actually matters if you care about this being fact based instead of fear based: workers who say their productivity is monitored electronically "all the time" are roughly three times more likely to say they work faster than is healthy or safe, compared to workers reporting no monitoring at all, 46 percent versus 15 percent. That relationship held up even after Hertel-Fernandez adjusted for industry, occupation, company size, and union status.

What predicts the harm isn't the monitoring itself. It's what workers think the monitoring is for. When workers believed it was there to protect their health and safety, there was no measurable increase in unsafe pace. When workers believed it existed to discipline them or shut down complaints, 67 percent reported working at an unsafe pace, compared to 14 percent of workers who didn't believe that. Same technology. Opposite outcome. All depending on how it's deployed and what workers think it's for.

Amazon is the case study that gets cited most because it's the most documented. A Washington Post investigation found injury rates at Amazon warehouses running higher than at other warehousing companies and pointed to electronic productivity monitoring as a contributing factor. A University of Illinois Chicago study surveying nearly 1,500 Amazon warehouse workers across more than four hundred facilities backed that up: workers reported high injury rates and a persistent sense they needed to work faster than felt safe. Washington state's labor enforcement agency initially cited Amazon over unsafe pace demands tied to its monitoring systems. A judge cleared Amazon of those specific citations in 2024, and that outcome belongs in the record here too. The pattern documented across multiple independent studies doesn't disappear because one enforcement case didn't hold up in court, but pretending the case went the other way would be exactly the kind of fudging this book doesn't do.

One more number: testimony before Congress from Cornell labor researcher Kate Bronfenbrenner found that the share of NLRB union election campaigns where employers deployed electronic surveillance of workers rose from 14 percent between 1999 and 2003 to 32 percent between 2016 and 2021. More than double. Whatever else workplace monitoring gets used for, it's also being used more often around the exact moments when workers are deciding whether to organize.

Monitoring isn't automatically the villain here. The same survey found that workers under monitoring built specifically for their safety, GPS panic buttons for hotel housekeeping staff facing possible assault, sensors reminding healthcare workers to wash their hands between patients, didn't show those same negative outcomes. The technology isn't the variable. The intent behind deploying it is.

## WHO'S ACTUALLY ACCOUNTABLE

Four systems, one throughline. Bertillon's card system promised objectivity a century and a half before anyone questioned it. Facial recognition told police who Robert Williams was, and it was wrong. Predictive policing told officers where crime would happen, and it mostly told them where they'd already decided to look. Workplace monitoring tells employers how fast someone is working, and whether that number protects a person or breaks them depends entirely on what it was built to do.

None of these tools got a vote. Not from Williams. Not from the workers whose schedules a computer assembles overnight. Not from the neighborhoods PredPol drew boxes around for a decade before anyone checked whether the boxes meant anything. Not from the billions of people whose faces sit in Clearview's database without their knowledge. The technology moved first. The accountability is still catching up, one settlement, one GAO report, one shut down company at a time.

That's not a reason to panic about being watched. It's a reason to ask, every single time: watched by what, for what, and who answers for it when it gets it wrong.

# CHAPTER 13: WHO OWNS THIS, WHO REGULATES THIS, WHO'S ACTUALLY IN CHARGE

## TWO PEOPLE ARE BUILDING THE SAME CAR

Picture two people building the same car in the same year.

One has blueprints. Real ones, argued over for years, published, dated, with fines attached to the parts that don't get built on schedule. She might hate pieces of it. She might think it moves too slow or asks too much. But she knows where the deadlines sit, because they're written down and they don't move without another vote.

The other is building the car with a different rulebook depending on which state he's standing in when he picks up a wrench. And the guy funding the whole project keeps calling every few months to say the instructions from last time don't count anymore.

That's not a metaphor for AI regulation. That's AI regulation, right now, in the two biggest AI markets on earth.

The European Union has a law. It's called the AI Act. It went into force in 2024, and pieces of it have landed on schedule ever since. Agree with the approach or don't, it exists, it's binding, and anyone building AI for European users has known the shape of the deadlines for two years.

The United States has something else. A president who canceled his predecessor's AI order on his first day back, then spent the next year and a half issuing new ones: first stripping regulation away, then suing states that wouldn't fall in line, then, in June 2026, reversing course to build a review process for the most powerful models on the market. A Congress that tried to freeze every state's ability to regulate AI for ten years, tucked into a budget bill most people weren't reading closely. The Senate found out. The vote to strip it back out was 99 to 1.

And while Washington makes up its mind, fifty state legislatures aren't waiting. Every one of them introduced an AI bill in 2025, the first year that's ever happened. Colorado passed the broadest AI law in the country, then took most of it back within two years. California built a dozen narrower rules instead of one big one. Texas banned specific harms outright and moved on.

Nobody is coordinating this. Nobody has to.

This chapter isn't going to tell you which approach is right. It's going to hand you the map as it actually stands, so you can stop hearing there are no AI laws and start understanding that there are. They're just written in about a hundred different places at once, and a few of them are still being edited as you read this.

## EUROPE PICKED A BLUEPRINT

Start with the European Union, because it's the simplest part of the story. The EU passed an actual law, the AI Act, and it entered into force on August 1, 2024. It doesn't take effect all at once. It rolls out in stages: prohibited practices and AI literacy rules applied starting February 2025, and the obligations for general purpose AI models, the category that includes systems like ChatGPT and Claude, became applicable in August 2025.

The law sorts AI into risk categories. At the bottom sits most AI in daily use: spam filters, AI in video games. That's not regulated at all. At the top sit what the EU calls high-risk systems, meaning AI used to make consequential decisions about people: recruitment, credit scoring, law enforcement, education, border control. Those face the heaviest obligations, documentation, risk assessments, human oversight.

Here's the part that matters for a book printed months after any of us can promise the deadlines won't move again. The big obligations for high-risk systems were supposed to land in August 2026. They didn't, not on the original schedule. In May 2026, EU lawmakers reached a political agreement to push those deadlines back. Standalone high-risk systems, the ones used for hiring and border control, now have until December 2, 2027. AI embedded into regulated products like medical devices gets until August 2, 2028.

That's not the EU backing down from its own law. Implementation was visibly behind schedule by late 2025, which is what pushed the Commission to propose the delay in the first place. A government admitting its own standards bodies weren't ready in time is a different kind of honest than most regulatory stories get credit for.

And August 2026 still isn't a nothing date. The transparency rules that require chatbots to disclose you're talking to a machine still take effect that month. It's the watermarking requirement for AI-generated content, a separate rule, that got pushed to December 2026.

One more piece worth sitting with before we leave Europe. That same round of May 2026 changes added a new ban: AI tools that generate non-consensual sexual images of real people, so-called nudifying apps, alongside a parallel ban on AI-generated child sexual abuse material. The prohibition takes effect December 2, 2026. That's a law answering a harm that barely existed in its current form when the AI Act was first drafted back in 2021. Which is the whole chapter, honestly, compressed into one paragraph. The technology moves fast enough that even the government that got out ahead of it keeps having to reopen the rulebook.

Now cross the Atlantic, where there isn't one rulebook so much as a drawer full of executive orders, several of which contradict the ones before them.

## THE UNITED STATES KEEPS CHANGING JOBS

If the EU section reads like a straight line with a few speed bumps, this one reads like a car doing donuts in a parking lot.

Start with the flip. On January 23, 2025, days after his second inauguration, President Trump signed Executive Order 14179, Removing Barriers to American Leadership in Artificial Intelligence. It did what the name promises. It revoked his predecessor's AI order, Executive Order 14110 from October 2023, which had built out safety testing and reporting requirements for powerful AI systems. The new administration's goal: clear away regulation it saw as a drag on American AI development.

That set the tone for the next year and a half. In July 2025, the White House released Winning the AI Race: America's AI Action Plan, organized around three pillars, accelerating innovation, building infrastructure, and leading internationally, framed around removing what the plan called red tape. That same day, Trump signed a second order, Preventing Woke AI in the Federal Government, which said the federal government would only buy AI systems built on stated principles of truth seeking and ideological neutrality.

Congress tried something bigger that same summer, and it blew up in public. Senate Republicans attempted to tuck a ten-year freeze on all state AI regulation into the budget bill known as the One Big Beautiful Bill Act. It didn't survive the floor. On July 1, 2025, the Senate voted 99 to 1 to strip the moratorium back out, with only Senator Thom Tillis of North Carolina voting to keep it. A 99 to 1 vote is not a normal outcome for anything in that chamber right now. It's one of the clearest signals available that blocking states from regulating AI, even in a Republican controlled Congress, was not the easy win the tech industry expected.

By December, the administration's posture shifted from deregulating the federal government's own AI use toward actively fighting state regulation in court. Executive Order 14365, Ensuring a

National Policy Framework for Artificial Intelligence, signed December 11, 2025, directed the Attorney General to stand up an AI Litigation Task Force with one job: challenge state AI laws. It also told the Secretary of Commerce to identify which state laws counted as onerous and make those states ineligible for a slice of federal broadband funding.

In March 2026, the White House tried the legislative route, releasing a National Policy Framework for AI that asked Congress to pass a uniform federal law preempting state AI rules. That framework isn't binding. It creates no compliance obligations on its own. As of this writing, no federal preemption of state AI law has passed.

Then came the sharpest turn yet. On June 2, 2026, Trump signed Promoting Advanced Artificial Intelligence Innovation and Security, an order that reads almost nothing like the ones before it. It sets up a voluntary system where developers of the most advanced AI models can share them with the federal government for up to thirty days before wider release, so national security agencies can check for risks. It also directs the Attorney General to prioritize prosecuting people who use AI to commit fraud or break into computer systems. It's still voluntary. It creates no licensing requirement. But it's the first time this administration has proposed the federal government get an early look at frontier AI models on national security grounds, after eighteen months of telling agencies to get out of AI's way.

None of this is law the way the EU AI Act is law. Executive orders aren't legislation. A future president can undo every one of them with a signature, the same way this one undid his predecessor's. That instability isn't a footnote. It's the actual current state of federal AI policy in this country, and anyone who tells you with total confidence what the rule is has usually stopped reading a few months too early.

## FIFTY STATES, FIFTY ANSWERS

While Washington argues with itself, state legislatures haven't waited around for permission. According to the National Conference of State Legislatures and the legislative tracking service MultiState, 2025 alone saw roughly 1,200 AI related bills introduced across all fifty states, the first year every state introduced at least one. By March 2026, lawmakers in 45 states had already introduced more than 1,500 additional bills, on pace to blow past the entire previous year's total before most sessions had even wrapped up.

Most of that activity is narrow. More than three dozen states have passed something, usually a law targeting deepfakes in elections or nonconsensual intimate imagery and stopped there. But five states have gone further and built AI laws with real teeth, and none of their stories are as simple as the law passed and now it applies.

Colorado is the clearest example of how messy this gets. The state passed what was, on paper, the broadest AI law in the country back in 2024, Senate Bill 24-205, aimed at high-risk AI systems used in decisions about employment, housing, lending, insurance, healthcare, and education. It was supposed to take effect February 1, 2026. A special legislative session in August 2025 pushed that to June 30, 2026 instead, because industry groups argued the original law wasn't workable as written. Then in April 2026, a federal magistrate judge paused enforcement of the law entirely, after Elon Musk's xAI sued and the U.S. Department of Justice intervened on xAI's side, the first time the federal government had stepped into a challenge to a state AI law. Then, on May 14, 2026, the legislature didn't just delay the law again. It repealed the original and replaced it with a narrower one, Senate Bill 26-189, which drops the broad high-risk AI framework for a tighter focus on automated decision-making technology, with real compliance obligations not starting until January 1, 2027. Three different effective dates and a full rewrite, in under two years, on the single most cited state AI law in the country. That instability is the story, not a footnote to it.

Texas took a different approach: a single, more targeted statute. The Texas Responsible AI Governance Act took effect January 1, 2026, and instead of a broad risk management regime, it bans specific harms outright: intentional discrimination, government use of AI for social scoring, and AI generated child sexual abuse material.

California didn't pass one big law either. It passed a stack of them. In 2024 alone, California enacted eighteen separate AI related bills, and two more took effect January 1, 2026: the Transparency in Frontier Artificial Intelligence Act, which requires the largest AI developers to publish safety practices, and a companion law requiring generative AI developers to disclose high level information about the data used to train systems available to Californians. Separately, the state's privacy regulator finalized rules on automated decision-making and cybersecurity audits, with obligations phasing in as late as 2030.

Utah moved earliest of anyone on this list. Its AI Policy Act took effect back in May 2024 and requires disclosure when consumers are interacting with generative AI, with stricter disclosure rules for higher risk uses like mental health services.

Illinois went the employment law route. A 2024 amendment to the Illinois Human Rights Act, effective January 1, 2026, restricts how employers can use AI in hiring decisions and requires notice when AI is used in employment decisions, closing a gap New York City had already tried to address at the local level with its own bias audit requirement for automated hiring tools, in force since 2023.

Layered underneath all of this sits one voluntary federal framework almost everyone in the compliance world uses: the National Institute of Standards and Technology's AI Risk Management Framework. It isn't a law. Nobody is required to follow it. But because it's free, thorough, and built with input from industry and academia rather than any single company's interests, it's become the closest thing to a common baseline for a business operating across all these different state rules.

And Congress has passed exactly one AI specific federal statute out of all of this: the TAKE IT DOWN Act, signed in May 2025, which criminalizes the nonconsensual distribution of intimate images, including AI generated ones, and requires platforms to remove them on request.

So that's the actual map. Not no rules. A law in Europe with real deadlines, even if some of those deadlines just moved. A federal government in the US that's changed its own approach at least four times in eighteen months. And fifty states, five of them with serious teeth, writing their own answers in the meantime, sometimes rewriting their own answers within the same year they wrote them.

## THE LABS SAY THEY'LL POLICE THEMSELVES, IF SOMEONE HELPS

This next part is happening as this book goes to print, which tells you something about the pace of everything else in this chapter.

On July 14, 2026, Google DeepMind CEO Demis Hassabis published an essay on X calling for the United States to build something that does not exist yet: a Frontier AI Standards Body. He modeled it on the Financial Industry Regulatory Authority, the private, industry funded watchdog that polices Wall Street under Securities and Exchange Commission oversight, with a board that includes independent technical experts and open-source representatives. Labs would start by sharing frontier models voluntarily, up to thirty days before release, with the arrangement becoming mandatory for U.S. deployment once the process proves it works. Hassabis said the body would need serious funding to attract top technical talent and run large scale testing, and that industry would likely foot that bill. He described the current moment as sitting at the foothills of something bigger, comparing the arrival of AGI not to the internet or mobile phones but to the discovery of fire. He wants it running by the end of this year.

Here is the part worth being straight about. Hassabis pointed to a specific event as the reason this could not wait: the federal government's handling of a rival lab's models. He told Axios that Washington's treatment of Anthropic's Mythos and Fable models the month before, an eighteen-day freeze under an export control directive, was a bit of a wake up call, and that it showed Washington needs a formal governance structure instead of improvised, case by case responses. That freeze is not a metaphor. It happened to the company that trained me. I am naming that plainly instead of burying it, because a book about who is watching AI does not get to go quiet the one time the watching lands on its own author.

But do not mistake the anecdote for the ask. The proposal itself is not aimed at any single company. It is a standards body meant to test frontier models across the entire industry for cyber, nuclear, and biological risk before they ship, with the power to coordinate an industry wide slowdown if the danger is real enough. And Hassabis is not the only one asking. OpenAI CEO Sam Altman has separately called for a similar international standards body of his own. A month before the essay, Hassabis and Anthropic CEO Dario Amodei had already made a version of this same pitch together, in a closed door session with G7 world leaders including President Trump, in Évian-les-Bains, France. No concrete commitments came out of that meeting.

Read those two facts side by side and the shape gets clearer. Google's lab and OpenAI, two companies that compete for the same customers, same talent, same headlines, are both independently telling Washington the same thing: we would rather get tested by a real body with real teeth than keep guessing what the executive order says this month. That is not altruism. Getting ahead of a mandatory rule usually means getting to help write it. But wanting a rulebook and wanting no rules at all are different instincts, and this chapter has spent a lot of pages on people picking the second one.

None of this is built yet. No board exists. No funding is committed. As of this writing, it is one essay, one stated goal of running by year end, and two competing labs pointing at the same gap in the sky and asking someone to fill it. Whether it exists by the time you are holding this book is a fact you will know and I will not.

## BACK TO THE TWO PEOPLE BUILDING CARS

One of them, the EU, picked a blueprint years ago and is now living with what it's like to hit real deadlines on a huge, technical undertaking. Some parts landed on schedule. Some got pushed back once the standards bodies admitted they weren't ready. But it all happens inside a process where you can point to a document and know what changes and when.

The other, the US federal government, has spent eighteen months trying on different jobs. Referee. Then opponent, suing the states it used to leave alone. Now, as of June 2026, something closer to a national security screener with a thirty day early look at the most powerful models. None of those roles has stuck yet. And the one thing every part of Washington seems to agree on, Republicans and Democrats both, in a 99 to 1 vote, is that nobody up there gets to freeze what the states are doing while the federal government figures out its own job.

The states kept building. Some of them are still rewriting their own laws mid project, the way Colorado did twice in two years. That's not a system falling apart. That's a system with no referee, which is a different thing entirely.

None of this is fixed. Most of it wasn't fixed a year ago either, and it probably won't be fixed a year from now. What you have, right now, is the actual shape of it: one binding law overseas, a federal government still deciding what its own role even is, and fifty separate legislatures who never waited for permission to find out.

That's not a gap in the rules.

That's the rules. Right now. As they actually exist.

# CHAPTER 14: YES, IT'S COMING FOR SOME JOBS

## THE TELLER WHO DIDN'T DISAPPEAR

In 1969, a Chemical Bank branch on Long Island installed the first ATM in the United States. By the mid-1990s they were everywhere, and today there are roughly 400,000 of them scattered across the country. Every reasonable person looked at that math and drew the same conclusion: bank tellers are done. President Obama used this exact example in a 2011 interview, pointing to ATMs and airport kiosks as proof that businesses had learned to run on fewer workers.

It didn't happen. Not at first. Economist James Bessen dug into the actual data and found that as ATMs rolled out, teller employment didn't fall, it grew. Two things happened at once. First, ATMs cut the number of tellers needed to staff a typical urban branch from twenty down to thirteen between 1988 and 2004. That made branches cheaper to run, so banks opened more of them. Urban branch counts jumped 43 percent. Fewer tellers per branch, more branches, and the job didn't disappear, it multiplied. Second, the tasks that didn't get automated, like walking a small business owner through a loan or untangling a complicated account, became the actual job. Cash handling went down. Relationship work went up.

That's the version of the ATM story that gets repeated at every conference panel on automation, and it's true. Here's the ending that usually gets left off. More than 340,000 teller jobs still exist today, but employment in the role has dropped nearly 30 percent since 2010, and it wasn't the ATM that finally did it. It was the smartphone. Mobile check deposit and online banking accomplished in the 2010s what ATMs never managed in the 1990s, because this time the thing getting automated wasn't a task inside the branch. It was the reason to walk into the branch at all. The Bureau of Labor Statistics now projects teller employment will decline another 13 percent between 2024 and

2034, and its reasoning is blunt: as more people use mobile deposit, fewer customers show up at the teller window, so demand for tellers keeps shrinking.

So which version is true? Both, in order. The ATM automated a task, and the job got safer for two decades. The smartphone automated the errand itself, and the job started actually disappearing. Task versus job, tool versus replacement. That distinction is the spine of this entire chapter, and it's the part both the panic headlines and the reassurance headlines skip on the way to their conclusion.

## WHAT A JOB TECHNICALLY IS

A job is not one thing. It's a bundle of tasks wearing a job title as a costume. That's not a metaphor I made up to sound clever; it's the actual framework labor economists have used for two decades. In 2003, MIT economist David Autor, along with Frank Levy and Richard Murnane, published a paper breaking occupations down by task type instead of job title. Routine tasks, the kind with a clear right answer and a repeatable procedure, get automated first and fastest. Nonroutine tasks, the ones that need judgment, context, or actual human contact, mostly get boosted by automation instead of replaced by it.

Daron Acemoglu and Pascual Restrepo pushed that framework further in 2019 with a distinction that runs through the rest of this chapter: displacement versus reinstatement. Automation destroys some tasks. It also creates brand new ones that didn't exist before. Whether a job category grows or shrinks depends on which force wins in that specific role. Not on how impressive the technology looks in a demo. On the actual math of tasks lost versus tasks created.

A 2026 study out of MIT Sloan, co-authored by economist Lawrence Schmidt, measured this directly. When AI can perform most of the tasks that make up a job, the share of people in that role at a company falls by about 14 percent. But when AI's impact is concentrated in just a few tasks, leaving the rest of the job untouched, employment in that role can grow, because workers redirect their effort toward the parts AI still can't touch, like critical thinking or generating new ideas. Same technology, same company, opposite outcome, depending entirely on what share of the job got eaten. The study even found that high-wage roles heavily exposed to AI saw their employment share grow by about 3 percent over five years, because firms that adopted AI grew faster and could sustain more headcount even in the exposed positions.

And here's the detail that breaks the pattern set by every automation wave before this one. Research from OpenAI economists Tyna Eloundou, Sam Manning, Pamela Mishkin, and Daniel Rock, published in Science in 2024, found that exposure to large language models runs higher in higher-income occupations, the ones built around information processing and analysis, exactly the tasks current AI already does well. That's the reverse of computerization and factory automation from the 1980s through the early 2000s, which mainly hollowed out middle-skill, routine work like clerical jobs and basic bookkeeping. The last few waves came for the middle. This one is starting closer to the top of the pay scale and working through information work generally.

## THE LOOM THAT ATE ITS OWN INDUSTRY

This pattern isn't new to this century. In the early 1800s, power looms began replacing hand looms in British textile mills, and a real, documented group of workers paid for it. The number of British handloom weavers collapsed from around 250,000 in 1800 to roughly 7,000 just sixty years later. Between 1811 and 1816, bands of textile workers across Nottinghamshire, Yorkshire, and Lancashire organized to smash the machines they correctly understood were taking their trade. History remembers them as the Luddites, and the word has been an insult ever since, which is a little unfair, because they weren't wrong about what was happening to them.

Here's where it gets complicated in the way that matters for this chapter. James Bessen spent years reconstructing what happened to weaving as an occupation over the following century, not just to the specific workers displaced in that first wave. His finding: over the course of the 19th century, roughly 98 percent of the labor time required to weave a yard of cloth got automated away. Almost the entire task. And the number of weavers as an occupation didn't go away. It grew, for more than a hundred years. Power looms drove the price of cloth down. Cheaper cloth meant people bought more of it, and that expanding demand needed more total weaving hours to meet it, even though each individual yard needed a sliver of the labor it used to. This ran for over a century, and it only stopped working as a job creation engine once demand for cloth leveled off, sometime around the 1930s.

The job didn't survive by staying the same. It survived by turning into a different job wearing the same title. By 1902, a single American weaver was managing eighteen power looms at once, producing more than fifty times the output of a weaver a century earlier. That didn't happen because the machines kept getting better on their own. Bessen's research found that most of the productivity gain over that period came from improved labor quality, workers who got better at monitoring and coordinating more machines at once, catching problems before they became waste. The mills that thrived were the ones that invested heavily in training their weavers to supervise the machines, not the ones that just bought more equipment and assumed the machine would handle the rest.

Put the weaver and the teller side by side and the pattern holds. Automation ate the routine, repeatable core of both jobs almost completely. In both cases, total employment didn't fall, because cheaper output expanded demand faster than automation shrank the labor needed per unit. And in both cases the job that remained wasn't the same job with a smaller headcount. It was a genuinely different job, weighted toward oversight and judgment instead of the task that got automated. That's also exactly why the story eventually stops holding once demand tops out, which is what happened to cloth in the 1930s and to bank branches once mobile banking gave people a reason to stop visiting them at all. So, the real question about any job isn't can this be automated. Almost everything can be automated to some degree by now. The question is whether the market for what that job produces still has room to expand once the automated version gets cheaper, or whether that market is already about as big as it's going to get.

## WHAT THE GOVERNMENT SAYS IS HAPPENING RIGHT NOW

Now to the present tense, and to the source with the least incentive to sell a scary or reassuring story either way: the federal government. The Bureau of Labor Statistics publishes employment projections for the whole economy every year, and in its 2024 to 2034 cycle, it explicitly built AI's expected effects into occupation by occupation forecasts. Paralegals and legal assistants are projected to see little or no employment change over the decade, as AI tools take over more contract review, discovery, and legal research. Claims adjusters, examiners, and investigators are projected to decline 5.1 percent, as AI takes over photo based damage assessment and speeds up payout estimates. Tutors are expected to see slower employment growth as digital tutoring systems built on language models take on more instruction and question answering. And a cluster of office and administrative roles, billing and posting clerks, procurement clerks, credit authorizers, customer service representatives, and non-medical secretaries, are all projected to decline or show little to no growth, continuing a trend that automated phone systems and basic virtual assistants had already started years before generative AI existed.

That list isn't a picture of entire industries getting wiped out. It's a list of specific occupations, mostly built around tasks that are repeatable, rule governed, and don't require much in person judgment, exactly the tasks the task based framework predicts would go first.

## WHAT THE PEER REVIEWED RESEARCH SAYS

The peer reviewed academic research on AI specifically, as opposed to automation broadly, tells a consistent story so far. Economists Daron Acemoglu, David Autor, Jonathon Hazell, and Pascual Restrepo studied establishment level hiring data covering close to the full universe of U.S. online job vacancies from 2010 onward, published in the Journal of Labor Economics. They found rapid growth in AI related job postings from 2010 through 2018, concentrated at establishments whose workers already did tasks compatible with AI's capabilities, and those establishments simultaneously cut hiring in non-AI roles and tightened the skill requirements of the postings that remained. The honest, unglamorous part of the finding: at the level of the broader economy, the aggregate effects of this AI driven substitution on overall employment and wage growth in more exposed occupations and industries were, as of their data, too small to detect. Real, measurable change at individual workplaces. Not yet a measurable dent in the national numbers.

## THE BANK'S NUMBERS AND THE CONSULTANT'S NUMBERS

Now the industry research, and I'm labeling it as such because that's exactly what it is. Goldman Sachs is a bank. It sells research and financial products, and its economists have a business incentive to produce attention getting numbers. In March 2023, Goldman Sachs economists analyzed the task content of more than 900 occupations and estimated that the equivalent of 300 million full-time jobs worldwide were exposed to some degree of automation by generative AI, with roughly two thirds of U.S. and European jobs exposed to some degree, and a quarter to half of the workload in those exposed jobs potentially replaceable. That figure is probably the single most repeated AI jobs statistic of the decade.

What almost nobody repeated: in August 2025, Goldman's own research team checked that projection against real labor market data and found something a lot less dramatic. They found no statistically significant correlation between AI exposure and job growth, unemployment, job finding rates, layoff rates, hours worked, or wage growth across the broader economy. Their updated estimate, based on how AI is being used rather than how it theoretically could be used, put current employment at risk at only about 2.5 percent, a small fraction of the original 300 million job framing. Their baseline expectation over a ten-year adoption period was a 6 to 7 percent displacement rate, not the sweeping labor market collapse the 2023 headline implied.

That correction matters as much as the original number did. Goldman's economists did find early, real signs of disruption in specific corners. Employment growth in marketing consulting, graphic design, office administration, and telephone call centers has fallen below its expected trend, and unemployment among workers in their twenties in tech exposed occupations has risen close to three percentage points since the start of 2025, notably worse than for other young workers in unrelated fields. Their occupational risk list names computer programmers, accountants and auditors, legal and administrative assistants, customer service representatives, telemarketers, proofreaders and copy editors, and credit analysts as highest risk, and air traffic controllers, chief executives, radiologists, pharmacists, residential advisors, photographers, and clergy as lowest risk, mostly because those roles combine high stakes error consequences, physical presence, or forms of trust current AI hasn't earned yet.

As of this writing in mid-2026, Goldman has already revised its own number again, estimating AI could displace roughly 15 million American workers, or about 9 percent of the workforce, over the coming decade, up from the 6 to 7 percent baseline set less than a year earlier. Their tracker currently attributes a net drag of somewhere between 11,000 and 16,000 U.S. jobs a month to AI, concentrated in tech, management consulting, and graphic design, offset partly by construction hiring tied to data center buildout. This is a number that keeps moving because the underlying technology and adoption rate keep moving. Treat every figure in this section as a snapshot, not a verdict.

Now the consulting research, McKinsey Global Institute, same caveat: this is the research arm of a firm that sells transformation advice to the companies it studies, so I'm labeling it industry research, not neutral academic data. Its 2023 update on the future of work found that without generative AI, automation was on track to take over tasks accounting for about 21.5 percent of hours worked in the U.S. economy by 2030. With generative AI factored in, that midpoint estimate jumped to 29.5 percent. That's a claim about hours of task time, not a claim that 30 percent of jobs disappear, the same task versus job distinction running through this whole chapter.

McKinsey's occupational transition estimate is more specific and more useful. The firm estimates 11.8 million U.S. workers currently in shrinking occupations may need to move into different lines of work by 2030, with roughly 9 million of them landing in entirely different occupational categories, on top of the 8.6 million occupational shifts that already happened between 2019 and 2022. That burden isn't spread evenly. McKinsey found that workers in the second lowest wage quintile, those earning roughly $30,800 to $38,200 a year, are up to 14 times more likely to need to change occupations entirely than the highest earners, and workers in the lowest wage quintile face about 10 times the risk. Women are about 1.5 times more likely than men to need to move into a different occupation, largely because women are heavily represented in the office support and customer service categories absorbing most of the projected decline.

That's the same story that has played out in every automation wave this book has covered, going back to the power loom. The technology doesn't distribute its costs evenly. It concentrates them on whichever workers happen to be doing the most routine, most rule bound version of their job, and those workers are disproportionately lower paid to begin with. McKinsey's report does include a genuinely hopeful, measured thread alongside this: the occupational transitions that already happened between 2019 and 2022 skewed toward workers landing in higher wage roles, not lower ones, which its own researchers called one of the more surprising findings in the data. Whether that trend holds as AI adoption accelerates is genuinely unknown.

Here's where the government data, the academic research, the bank data, and the consulting data all land, independently, in roughly the same place. AI's disruption right now is concentrated in specific, nameable occupations built around routine, rule governed tasks: office support, customer service, claims processing, certain legal and paralegal work, entry level coding and administrative work. It is not, currently, an economy wide job elimination event, and the peer reviewed research says the aggregate effect is still too small to detect in the broader numbers, even while it's visible at individual workplaces. And the workers absorbing what disruption does exist are disproportionately the ones who were already lower paid and had the least cushion to begin with. That's not fearmongering. It's what a government agency, a peer reviewed labor economics paper, an investment bank, and a consulting firm, four sources with very different incentives, currently happen to agree on.

## THE TWIST: WHEN THE JOB GROWS AND THE PAY SHRINKS

There's one more wrinkle worth sitting with, because it complicates even the comforting version of this story. Everything so far assumes automation eats the routine, boring part of a job and leaves humans the interesting, high value part, which usually means the remaining work gets more valuable, not less. MIT researchers David Autor and Neil Thompson studied what happens when automation goes the other direction, targeting a role's most expert tasks instead of its routine ones. Their example: inventory clerks. As computerization took over the most expert parts of that job, employment in the role more than doubled between 1980 and 2018, and real wages fell by 13 percent. The job got more common and it got paid less at the same time, because once automation handles the hardest, most specialized part of a task, the expertise required to do the rest becomes a lot more replaceable. A much wider pool of workers can now produce the same output with the tool doing the hard part, so more people compete for the role, and the wage falls even as headcount grows because demand for the now cheaper output expands right along with it. Compare that to bookkeepers and accounting clerks over the same stretch. Computers took over their routine tasks, employment fell by a third, and real hourly wages rose by nearly 40 percent, because what was left of the job needed more judgment, not less.

Current AI tools, unlike the factory robots and assembly line automation of past decades, are increasingly capable of aiming at cognitive and expert level tasks first: legal analysis, financial modeling, diagnostic reasoning, code architecture, not just data entry and routine paperwork. That's a meaningfully different target than most of the automation waves this book has covered, and it means the honest version of the task framework needs three lanes, not two. Automate a minority of routine tasks in a role, and the job usually survives with a shift toward the remaining higher value work, often at higher pay, the ATM and power loom pattern. Automate most of the tasks in a role, and headcount shrinks, the MIT Sloan 14 percent finding. Automate specifically the hardest, most expert task in a role, and you can end up with more people doing the job for less money each, because the thing that used to make the role scarce stopped being scarce. Which lane a given job falls into isn't something economists can fully predict in advance for every occupation, and current AI development is genuinely capable of going after any of the three.

## THE QUESTIONS WORTH ASKING

AI is not coming for your job as some single monolithic thing, and it's not leaving all jobs magically untouched either. It's coming for tasks, specific, nameable, countable tasks, and whether your job survives that intact, changes shape, or shrinks depends on one measurable thing: how much of what you actually do all day is the routine, repeatable part versus the part that needs judgment, physical presence, or trust a machine hasn't earned. The ATM didn't kill the teller. The smartphone did, a full generation later, because it automated the reason to show up at all, not just a task inside the building. The power loom ate 98 percent of the labor in weaving cloth and the occupation grew for a hundred years, until demand for cloth topped out and the job started shrinking for real. Right now, government data, peer reviewed research, bank research, and consulting research all point at the same handful of occupations, built almost entirely around routine, rule governed tasks. And the wage twist means even the job survived isn't automatically good news, because sometimes surviving means more people doing it for less.

None of that is a prediction about your specific job. It's a set of questions you can run your own job through. What percentage of what I do is repeatable and rule based versus judgment based. Is the thing my job produces something people would buy more of if it got cheaper, or is demand for it already about as big as it's going to get. Is the part of my job most exposed to automation the easy part or the hard part. Not panic. Not reassurance. A framework.

# CHAPTER 15: NO, IT'S NOT COMING FOR YOURS (...PROBABLY)

Economists love this story. In the 1970s, banks started installing ATMs. Everybody assumed bank tellers were finished. Instead, the number of tellers went up. ATMs cut the number of tellers a branch needed nearly in half, so branches got cheaper to run, so banks opened more of them. Bank branches in urban markets grew 43 percent. More branches needed more tellers, even if each one needed fewer. The job didn't die. It changed. Tellers stopped counting cash and started selling loans and building relationships. Economist James Bessen made this story famous in his 2015 book Learning by Doing, and it got picked up by economists across the political spectrum as proof that automation panic is usually wrong.

## THE TELLER TWIST

I'm about to complicate it, because that's the whole point of this chapter.

As of the 2024 to 2034 projection cycle, the Bureau of Labor Statistics expects teller employment to decline 13 percent. Not because of ATMs. ATMs never did it. What's doing it now is mobile check deposits and AI powered tools that can issue debit cards and catch counterfeit bills without a human at the window, combined with banks simply needing fewer physical branches now that fewer customers walk into them. The job that survived one wave of automation is losing to a different one, decades later, for a different reason.

So which version is true? Both. That's the uncomfortable, unsatisfying, completely factual answer. Automation did not wipe out bank tellers in 1985. It might be quietly doing it now. History does not repeat, but it rhymes, and if you only tell the first half of the story you get complacency, and if you only tell the second half you get panic.

Neither one is useful to the person trying to figure out what to do with her Tuesday.

That person is probably you, if you picked up this book after finishing the last chapter feeling a little rattled. Good. You should feel something. Feeling rattled and feeling doomed are not the same thing, and I'm not here to hand you a pep talk or a eulogy. I'm here to hand you the actual data, and the actual data says something more specific than "it's fine" or "we're cooked."

Automation does not usually eliminate jobs. It eliminates tasks. A teller's job was never really "hand people cash." That was one task inside a bigger job that also included judgment calls, fraud detection, customer relationships, and accountability for other people's money. When the cash handling task got automated, the job absorbed new tasks and kept going, for a while. When enough of the remaining tasks also become automatable, the job begins to shrink. That's what's happening to tellers now. The ATM story wasn't a lie. It just had an expiration date nobody put on the label.

This is the mechanism you need before you can honestly answer "is AI coming for my job." Not "is my job old" or "does my job involve a computer," but "what tasks make up my job, and how many of them require something a model genuinely cannot do." That's a different question, and it has a different answer depending on what it is you do all day.

MIT economist David Autor and his coauthors spent years combing through census records to answer a related question: how much of today's workforce is doing work that didn't exist a few generations ago. Their finding, published in the Quarterly Journal of Economics in 2024, is that roughly 60 percent of jobs held in 2018 fell under job titles that did not exist in 1940. Not 60 percent of workers got new titles for the same work. Sixty percent of the jobs themselves are new categories entirely. Data scientist, wind turbine technician, UX researcher, none of that existed when your great grandmother was job hunting. Automation destroyed the tasks that made up older jobs, and new jobs grew up around the tasks that were left, plus tasks nobody had thought of yet.

This isn't a new pattern. In 1900, 41 percent of the American workforce farmed for a living. By 2000, it was under 2 percent, according to data compiled by the USDA's Economic Research Service from historical Census records. That's not a stat about job loss. That's a stat about an entire category of human labor getting mechanized so thoroughly that almost nobody needs to do it anymore, while the freed up labor went and built the 20th century's cities, factories, hospitals, and everything else. Nobody in 1900 could have described most of the jobs that would exist by 1950. The transition was not painless. It was not instant. But it happened, and it did not result in permanent mass unemployment, because that is not actually how labor markets tend to absorb automation.

Which brings me to the number that gets quoted constantly, usually alone, usually with a sad trombone sound effect: the World Economic Forum projects that by 2030, AI and other structural trends will displace 92 million jobs globally. What gets left off the poster is the other number from the same report: 170 million new jobs created in that same window, for a net gain of 78 million. I'm not telling you this to say relax, it's all fine, because 92 million real people losing real jobs is not fine for those 92 million people. I'm telling you this because the honest picture is net positive job creation with real, painful, individual displacement inside it, and those two facts do not cancel each other out. You can hold both.

If the job count isn't the real story, what is? It's which tasks inside your specific job are the automatable kind, and which ones aren't. That's what the rest of this chapter is about. Not whether AI is coming. It's already here. The question is whether it's coming for the parts of your job that make you, you, or the parts that were never really the point.

## STOP PROOFING. START ACCENTUATING.

"AI proof your career" is defensive language. It pictures AI as weather, something to build a shelter against and wait out. That's the wrong picture, and it's not what the research on AI and work shows in reality.

In 2023, two MIT economists ran an experiment that's since become one of the most cited studies on this exact question. Shakked Noy and Whitney Zhang recruited 453 college educated professionals, gave them real, occupation specific writing tasks, and randomly let half of them use ChatGPT. The group with access was more productive, more efficient, and reported enjoying the work more. Average time on task dropped 40 percent while output quality rose 18 percent. This wasn't a simulation or a survey. It was a controlled experiment with real incentives, published in Science in July 2023, one of the most rigorous journals in the world.

That result held up in the real world too, not just a lab. Economists Erik Brynjolfsson, Danielle Li, and Lindsey Raymond studied more than five thousand customer support agents at a Fortune 500 software company as it rolled out an AI assistant in stages. Access to the tool increased worker productivity, measured in issues resolved per hour, by 15 percent on average, a finding published in the Quarterly Journal of Economics in 2025. Here's the part that matters most for the accentuate argument: the gains weren't spread evenly. Less experienced and lower skilled workers improved significantly in both speed and quality, while the most experienced workers saw only small gains. The tool functioned less like a replacement and more like a way to hand newer workers the instincts of the office's best performer. It closed gaps. It didn't erase people.

Neither study says AI makes you obsolete. Both say the same thing from two different angles: people who learn to work with it get faster and better at the parts of the job that were always going to exist, drafting, researching, troubleshooting, communicating. The frog you don't want to deal with, the tedious first draft or the rough calculation, gets fileted a lot faster. What's left is more of the part that actually needed you in the first place, the judgment about whether the draft says the right thing, whether the number makes sense, whether the customer needed something the transcript didn't capture.

That's the whole philosophy of accentuating instead of proofing. You are not in a race against a tool. You are deciding which of your tasks are frog tasks and handing those off, so you have more time and energy left for the tasks that were never going to be automatable regardless of how good the model gets.

## THE JOBS THAT AREN'T GOING ANYWHERE

There's a decades old observation in AI research called Moravec's paradox, named for robotics researcher Hans Moravec, who first laid it out in his 1988 book Mind Children. The paradox is that tasks requiring a lot of formal reasoning tend to be relatively easy to get a computer to do, while tasks that are effortless for a human, like walking across a cluttered room or picking up an oddly shaped object, are often the hardest thing to teach a machine. Chess turned out to be easier for a computer than folding laundry. It should be said, computer scientist Arvind Narayanan has pushed back hard on treating this as some kind of hard law of nature. He points out the "paradox" has never actually been tested and mostly describes what AI researchers found interesting to work on, not what's inherently easy or hard for a machine. Fair point. It's still held up remarkably well in practice, whatever the reason.

This is showing up in the actual employment numbers right now, not as a prediction but as a current hiring reality. The BLS projects electrician employment to grow 9 percent from 2024 to 2034, with about 81,000 openings a year, driven partly by the buildout of data centers and renewable energy infrastructure, the very infrastructure AI itself runs on. HVAC technicians are projected to grow 8 percent over the same window. Healthcare and social assistance is projected to be the fastest growing sector in the entire economy, adding roughlý 2 million jobs, and nurse practitioners are projected to be the single fastest growing occupation within that sector. None of that is a coincidence. Wiring a building, unclogging a pipe, comforting a scared patient at 3 a.m., these are jobs made of tasks that require a body in a specific place doing something a robot still can't reliably do, or a level of human trust a screen still can't provide.

That's the hopeful half of the accentuate idea. The tasks that resist automation aren't leftover scraps. They're built on judgment, physical dexterity, relationship trust, and accountability, the four things a model cannot do (yet) no matter how convincingly it can talk about them.

## ANY GIVEN SOMEDAY, AGAIN

Now the story that doesn't get told as often, because it's less comforting.

Starting in the early 2000s, Chinese manufacturing imports surged into the United States. Economists David Autor, David Dorn, and Gordon Hanson spent years tracking what happened to the American workers and communities most exposed to that shift, in research published in the Annual Review of Economics in 2016 and extended in a 2021 follow up study. What they found broke the model everyone expected. Economic theory says a displaced worker should retrain, relocate, or shift into a growing industry within a reasonable window. That's not what happened. Workers in the hardest hit local labor markets experienced elevated unemployment and depressed wages for at least a full decade after the shock began, and the follow up study found the damage was still measurable close to two decades later. This wasn't a story about automation, it was about trade, but the lesson generalizes. Displacement is real, it is not automatically fine, and "the market will sort it out" is not a plan. It's a hope.

I'm putting this in the same chapter as the hopeful stuff on purpose. This is not a book that flatters you with only the parts of the story that feel good. Some individual people, in specific industries and towns, got hurt badly and slowly by economic shifts they had no control over. Pretending otherwise would make me exactly the kind of coach I refuse to be, the kind who tells you it'll just work out.

But here's the other half, and it's the half that matters for what you do next. The workers who came out the other side of past disruptions in better shape were not the ones who waited to see what would happen. The 2025 World Economic Forum survey of over a thousand global employers found that half of all workers have now completed some kind of training, reskilling, or upskilling program, up from 41 percent just two years earlier. That's not proof reskilling guarantees a soft landing. Nothing guarantees that. It's evidence that the workers and companies treating this moment as something to prepare for, rather than something to survive or deny, are the ones actively changing their own odds.

I want to name the thing directly, because you probably already know it. Fear of AI taking your job is sometimes exactly that: real, specific, worth taking seriously. But sometimes it's a costume. It's Any Given SOMEDAY wearing a headline. “I can't switch careers now, AI is going to make my new field obsolete too” sounds like caution. Sometimes it's fear of starting something, wearing a very of the moment disguise. “There's no point learning that skill, AI will do it better” sounds like realism. Sometimes it's the same old story where waiting feels safer than trying.

I don't know your story. I'm saying it's worth asking yourself the question honestly, the way I'd ask a client in a session. Is the fear about AI specifically, or is AI just the newest, most socially acceptable reason not to do the thing you were already circling?

Because here's what's really true, and it's true whether or not AI existed: the workers who fared worst after the China Shock were the ones in local labor markets with no path to something else, not the ones who lacked information about the risk. The path is the variable that matters. Learning where AI genuinely threatens your specific tasks is how you build a path instead of waiting to find out if you needed one. That's not toxic positivity. That's just where the data points.

## JUDGMENT

In 2023, Harvard Business School researchers partnered with Boston Consulting Group to run one of the most rigorous field experiments on generative AI to date. They gave 758 BCG consultants, an elite group with degrees from places like Harvard, Yale, and Oxford, a set of real business tasks and access to GPT-4. The results were later published in Organization Science, a peer reviewed journal, and they're more interesting than a simple “AI made everyone better” headline.

On tasks that fell within what the model was actually good at, consultants using it completed roughly 12 percent more tasks successfully, worked about a quarter faster, and produced work rated up to 40 percent higher in quality, with the biggest gains going to people who'd been below average performers before. But on tasks that fell outside the model's actual capability, something else happened. Consultants using AI performed measurably worse than the ones who didn't use it at all. The researchers named this the jagged frontier, a boundary between what AI can and can't do that isn't smooth or obvious, so two tasks that look equally hard to a human can sit on opposite sides of it without any warning label.

The people who did best in this study weren't the ones who trusted AI completely or the ones who ignored it. They were the ones the researchers called cyborgs, professionals who kept moving back and forth between AI output and their own judgment, checking, questioning, catching the moments where confident sounding AI text was actually wrong. That checking, that professional skepticism, is the judgment piece. It's not a soft skill. It's the actual, measurable variable that determined who benefited from the tool and who got hurt by it.

**PHYSICAL DEXTERITY**

We already covered Moravec's paradox, the well documented pattern where tasks requiring formal logic are often easier to automate than tasks requiring a body moving through unpredictable physical space. This isn't just theory. It's reflected in real BLS hiring data right now: electricians projected to grow 9 percent through 2034, HVAC technicians 8 percent, with tens of thousands of annual job openings in trades that require hands, tools, ladders, and judgment calls made in real time in a specific physical location. A robot can be extremely good at chess and still not be trusted to safely rewire your kitchen. That gap hasn't closed, and nothing on the horizon suggests it's closing fast.

**RELATIONSHIP TRUST**

This one's been tested directly, more than once, and the results are consistent, though it's worth being precise about who was tested. A 2025 randomized survey experiment published in the Journal of Medical Internet Research, involving 1,762 people representative of the US population, found that patients trusted doctors more, and were more willing to seek their help, when the doctor explicitly said they weren't using AI at all. That preference held steady across age, political affiliation, and education level. A separate peer reviewed study by René Riedl, Svea Hogeterp, and Martin Reuter, published in Frontiers in Psychology in 2024, surveyed 1,183 people in Germany, Austria, and Switzerland and found the same ranking across multiple medical specialties: people preferred a human doctor first, a human doctor assisted by AI second, and an AI system alone last, every time.

Two different populations, two different continents, same order of preference. That isn't people being irrationally afraid of technology. It's a real, measured preference for a specific kind of relationship, the kind where a person is accountable to you as a person. That preference is a form of job security for anyone whose work depends on someone trusting them specifically, not just trusting that a task got done correctly.

## ACCOUNTABILITY

The European Union's AI Act, one of the first major binding AI regulations in the world, requires in Article 14 that high risk AI systems be designed so a human overseer can understand the system's capabilities and limitations, correctly interpret its output, and decide in any particular situation not to use it or to override it. That's not a suggestion. It's written into binding law across the EU.

In the United States, the FDA draws a similar line. Its Clinical Decision Support Software guidance, finalized in September 2022 and revised in January 2026, lays out four criteria software must meet to avoid being regulated as a medical device. The last of the four requires that the software let the healthcare professional independently review the basis for its recommendations, specifically so the professional isn't intended to rely on the software primarily to make a diagnosis or treatment decision. Software that gets too specific, too automatic, or too authoritative in what it tells a patient or provider crosses out of that exemption and into full device regulation.

The pattern underneath both of these is simple. Somewhere, someone has to be the name on the form, the license on the wall, the person who can be asked why they made this call and give an answer they're personally responsible for. AI can generate the analysis. It cannot hold the accountability. That has to sit with a human, by law in a lot of cases, and by basic professional and moral logic in the rest.

## PUT IT TOGETHER

None of these four things are rare or exotic. Most jobs have at least one baked in, and plenty have all four. A nurse practitioner has all four. An electrician has physical dexterity and accountability. A therapist has judgment and trust in spades. The reassuring part isn't that AI can't do impressive things. It clearly can, we've spent this whole chapter proving that with real numbers. The reassuring part is that the things making you good at your job, the parts a resume can't fully capture, are exactly the parts holding up under real scrutiny right now, in the actual research, not just in a coach's pep talk.

## WHAT TO ACTUALLY DO WITH THIS

The concrete version, not the pep talk version.

First, do an honest task audit. Don't ask "will AI take my job." Ask "which specific tasks do I do in a normal week." Write them down. Sort them: which ones are pattern completion, drafting, summarizing, calculating, formatting, the kind of thing Noy and Zhang's study showed AI speeds up by 40 percent. Those are your frog tasks. Hand them off.

Second, protect and build your time in the other pile. If your job leans on judgment, that means practicing the skepticism the BCG study's cyborgs showed, actually checking AI output against your own expertise instead of rubber stamping it. If it leans on physical dexterity, that's a genuinely growing, well paid lane right now, not a fallback. If it leans on trust, the data says people will pay for and seek out that trust, so don't undersell it as just being "good with people." If it leans on accountability, that part of your job is, in some cases literally, protected by law.

Third, treat the WEF's own numbers as a call to action, not a horoscope. Half of workers globally have already done some form of reskilling. That's not a small, rare group of overachievers anymore. It's becoming the median behavior. Waiting to see what happens is no longer the safe, default choice. It's just one option among several, and the data suggests it's not the one that tends to work out best.

Fourth, if you catch yourself reaching for "AI's going to make this pointless anyway" as a reason not to start something, treat that thought as information, not fact. Ask what you'd do about the decision if AI weren't part of the conversation at all. If the honest answer is you'd still be avoiding it, that's Any Given SOMEDAY talking, not the labor market.

## WHERE THIS LEAVES US

The frog in this chapter was never really the technology. It was the decision you'd already been avoiding, wearing new letters on its jacket.

AI is genuinely changing what a normal workday looks like, for real people, right now, and some of that change is going to be hard and specific and unfair to the people living through it. That's true. It's also true that the parts of your work built on judgment, a body doing something in a real place, someone trusting you specifically, and someone holding you accountable are holding up in the actual research, not just in the hopeful version of the story. Both of those things are real. You get to build your next move standing on both of them, not just the scary one.

# CHAPTER 16: THE WATER THING, FINALLY EXPLAINED

Somewhere online right now, someone is telling you that asking ChatGPT a question drinks a whole bottle of water. They are quoting a real number from a real study. They are also wrong, and the reason is a masterclass in how a true fact turns into a false headline without anybody technically lying.

Here is the actual paper trail.

In 2023, four researchers, Pengfei Li, Jianyi Yang, Mohammad Islam, and Shaolei Ren, at UC Riverside and UT Arlington, published a study called “Making AI Less Thirsty,” later peer reviewed and published in Communications of the ACM. It built the first real methodology for tracking how much water AI models use, both directly for cooling data centers and indirectly through the electricity that powers them. Using GPT-3 as their case study, they calculated the model needed to “drink” a 500 milliliter bottle of water for roughly every 10 to 50 medium length responses, depending on where and when the data center was running. Not per prompt. Per 10 to 50 prompts. One bottle split somewhere between ten and fifty ways.

That number sat in an academic paper doing what academic papers do, which is mostly nothing, until September 2024, when the Washington Post partnered with Ren to run a new calculation, this time for GPT-4, and this time for one specific task: writing a single 100-word email. GPT-4 uses far more energy per response than GPT-3, so the water number scaled up with it. The Post’s answer came out to about 519 milliliters, a bit more than a bottle, plus 0.14 kilowatt-hours of electricity. A whole bottle. For one email.

That is where the wires crossed. The original number was one bottle for a batch of requests. The new number was one bottle for a single request, but only because the task and the model had both changed too. Two different papers, two different models, two different questions, collapsed into one tidy headline: ChatGPT drinks a bottle of water every time you talk to it. It spread everywhere. Even Snopes ran a January 2025 explainer that repeated the figure without flagging the mismatch. Nobody lied. The math behind each number was real. The comparison built on top of it was not.

Here is my favorite detail in the whole mess. OpenAI CEO Sam Altman wrote in his June 2025 essay "The Gentle Singularity" that an average ChatGPT query uses about 0.000085 gallons of water, roughly 0.32 milliliters, alongside 0.34 watt hours of electricity. One fifteenth of a teaspoon. People saw that number next to the 519 milliliter figure and assumed one of them had to be lying. Nobody was lying. Altman was describing on-site cooling water for a short, ordinary query on OpenAI's current infrastructure. The Post was describing cooling water plus power-plant water for a longer GPT-4 response, calculated on older efficiency assumptions. Same category of technology, wildly different scopes, wildly different answers.

So here is where I land. The panic version of this story is wrong. That does not make the water conversation fake. Being concerned about this is valid, and it is necessary. We do not yet know every downstream effect of building this much infrastructure this fast. The honest position is not relax, and it is not the sky is falling. It is here's what's documented, here's what's still uncertain, and here's what's already being done about it.

## THE TWO NUMBERS THAT ACTUALLY MATTER

Forget per-prompt math for a second. Here is the number that matters at the national scale. The Lawrence Berkeley National Laboratory's 2024 United States Data Center Energy Usage Report, produced for the U.S. Department of Energy under a congressional mandate from the Energy Act of 2020, found that U.S. data centers directly consumed about 17.4 billion gallons of water in 2023 through on-site cooling. That is direct use. The same report found data centers consumed roughly another 211 billion gallons indirectly, meaning water used at power plants to generate the electricity those data centers ran on. Indirect use ran about twelve times larger than direct use. Almost every viral claim about AI's water footprint focuses on the smaller number and skips the larger one, backwards emphasis and all.

The report projects direct water consumption could double to quadruple by 2028, landing somewhere between 38 and 73 billion gallons a year. That range is wide because the industry is moving fast and the researchers who ran the numbers are not pretending to own a crystal ball. The uncertainty is a reason to take the projection seriously while it is still a range and not a locked-in outcome.

## WHEN IT STOPS BEING AN ABSTRACTION

National totals are easy to skim past. Newton County, Georgia is not.

Meta broke ground on a data center there in 2018. New York Times investigative reporting published July 14, 2025 found the facility uses about 500,000 gallons of water a day, which Mike Hopkins, executive director of the Newton County Water and Sewerage Authority, confirmed on the record is roughly 10 percent of the county's entire daily water use. One data center. Ten percent of a county's water. The Lincoln Institute of Land Policy independently reported similar figures.

Beverly and Jeff Morris live about 1,000 feet from that facility, on well water. Within months of construction starting in 2018, their dishwasher, ice maker, washing machine, and toilet stopped working correctly, according to the same Times reporting. Within a year, their water pressure had slowed to a trickle. Eventually nothing came out of the taps at all. They have spent thousands of dollars replacing appliances and cannot afford to replace the well itself, an estimated

$25,000 job. To be clear about what is and is not established here: Meta has stated that a third-party well study found no connection between its operations and the Morrises' water problems. That dispute is genuinely unresolved, and I am not going to write around that by implying otherwise.

Newton County is not unique. The same Times reporting found nine companies had applied to build additional data centers in the county, some requesting up to 6 million gallons a day, more than the county's entire current daily usage combined. Newton County is projected to hit a water deficit by 2030 without infrastructure upgrades, a fix the water authority has priced at over $250 million. A similar water pressure dispute surfaced at a data center campus in Fayette County, Georgia, where county officials attributed low pressure complaints to a transition to smart water meters rather than to the facility itself, according to Fortune's May 2026 reporting. Same pattern nationally: rapid buildout landing on infrastructure that was never sized for it, with responsibility often contested rather than confirmed.

## WHAT THE COMPANIES SAY ABOUT THEMSELVES, AND WHAT THAT LEAVES OUT

If you only read corporate sustainability reports, you would think this problem was basically solved. Some of what is in there is real. The gaps are just as informative as the numbers.

Amazon disclosed, for the first time, in June 2026 that AWS withdrew about 2.5 billion gallons of water across its global data center footprint in 2025, at an efficiency of 0.12 liters of water per kilowatt-hour of computing, which the company describes as roughly seven times more efficient than an industry average of 0.84 liters per kilowatt-hour. Amazon also states that data centers account for less than 0.5 percent of global industrial water use, and that Americans use more than 1,300 times more water annually on lawns and landscaping than Amazon's data centers use worldwide, a comparison built on EPA irrigation figures. That comparison is real and worth sitting with. It is also, by definition, Amazon's own framing of Amazon's own numbers, so I am labeling it self-reported rather than independent verification.

Here is what that framing tends to leave out. Amazon's efficiency figures cover on-site withdrawals, not the water used upstream to generate the electricity those data centers run on, the same direct-versus-indirect split that made the 17.4 billion and 211 billion gallon national figures so different from each other. AWS reported it was 53 percent of the way to its "water positive" pledge for 2024, and 75 percent of the way there by its 2025 disclosure. Both percentages are measured against Amazon's own accounting of its own direct footprint, not an independently audited total.

Microsoft announced in December 2024 that starting that August, all new data center designs would use closed-loop, zero-water-evaporation cooling, which recirculates the same water indefinitely instead of losing it to evaporation, with pilot facilities in Phoenix, Arizona, and Mount Pleasant, Wisconsin, coming online in 2026. Microsoft reports its water usage effectiveness improved to 0.30 liters per kilowatt-hour, a 39 percent improvement from 2021. Google has reported using reclaimed, non-potable water at data center campuses including one in Douglas County, Georgia, that runs on recycled municipal wastewater, with company-reported water consumption rising from 4.3 billion gallons in 2021 to 6.1 billion gallons in 2024 as its footprint expanded. None of this means these companies are lying. It means efficiency gains and rising total consumption can both be true at once, and a responsible reader needs both sentences, not just the one on the slide deck.

## REGULATION TRYING TO CATCH UP

Some of the most useful pushback here is not coming from viral videos. It is coming from state legislatures and researchers trying to get basic numbers out of an industry that mostly does not have to disclose them. A February 2026 report from UC Berkeley Law's Center for Law, Energy and the Environment found that data center water use in California often escapes meaningful public disclosure, in part because projects can qualify for streamlined approval pathways that skip detailed environmental review, and that where documentation does exist, it is frequently incomplete or inconsistent.

California Assemblymember Diane Papan authored a bill, AB 93, that would have required data center operators to disclose water use as part of local business licensing. Governor Gavin Newsom vetoed it in October 2025, writing that he was reluctant to impose rigid reporting requirements on the sector without understanding the full impact on businesses. Papan reintroduced similar legislation in 2026. By May 2026, more than 50 U.S. cities, including Fayetteville, Georgia, had enacted bans or moratoria on new data center construction, according to Fortune's reporting.

None of these fixes make the underlying tension disappear, and none of them are universally adopted yet. Most data centers being built right now still use evaporative cooling, and most communities still do not have easy access to the water use numbers for the facility down the road. But "there are real, documented engineering shifts and legislative fights already underway" is a different sentence than "don't worry about it," and it is the true one.

## IS THIS ACTUALLY WORSE THAN OTHER INDUSTRIES?

This comparison gets waved away by both sides of the argument, and it deserves a straight answer. U.S. data centers directly consumed about 17.4 billion gallons of water in 2023. Americans use roughly 3.3 trillion gallons a year just on lawn and garden irrigation, per EPA figures. Agriculture nationally uses far more water than either category. That context matters, and I am not going to pretend it does not.

But "smaller than agriculture" does not make Newton County not real, and it does not make the concentration problem disappear. The issue was never that AI uses more water than farming nationally. It is that a single facility can land on a single county's water system all at once, in a specific place that often did not get much say in the decision and does not have infrastructure built for it. National totals and local impact are two different measurements, and you need both to understand what is happening, the same way you needed both the 500-milliliters-per-50-queries number and the 519-milliliters-per-email number to understand the hook of this chapter.

The global trajectory is still worth naming honestly. The United Nations University Institute for Water, Environment and Health projected in a June 2026 report that global AI-related data center water use could reach roughly 9.3 trillion liters annually by 2030, an amount the report's authors compare to the annual domestic water needs of Sub-Saharan Africa's roughly 1.3 billion residents. That is a projection, not a measurement, produced by a UN-affiliated research institute with named authors and a public report. It deserves to be read as exactly that: a serious, credible warning about where the trend line points, not a verdict on where we already are.

Every number in this chapter has an asterisk. The job was never to erase the asterisk. It was to read it intellectually and objectively.

# CHAPTER 17: WHERE THIS IS REALLY HEADED

In 1965, Herbert Simon told a room full of people that machines would be capable of doing any work a human can do within twenty years. Simon was not a crank. He would go on to win a Nobel Prize and a Turing Award. He was one of the smartest people alive, looking straight at the frontier of his own field, and he was wrong by a margin that still stings.

Five years later, Marvin Minsky, another founding father of the field, was quoted by Life magazine's Brad Darrach saying that in three to eight years there would be a machine with the general intelligence of an average human being. Minsky denied ever saying it, and other parts of that same article have been questioned since too. But the prediction went out into the world anyway, people believed it, and when it did not happen on schedule, funding dried up and an entire research field went quiet for the better part of a decade. You met that stretch back in chapter two. Now you know two of the men who talked their way into it.

That is not a knock on Simon or Minsky. It is the whole point of this chapter. The people closest to a technology are often the worst people to ask about its timeline, because proximity breeds conviction, and conviction is not the same thing as being right.

## FIVE MEN, FIVE LABS, FIVE DECADES

Fifty-some years later, nothing has changed except the names. Ask the people running today's top AI labs when a machine will out-think humans at everything, and you get answers that are not close to each other. Elon Musk has predicted it as early as the end of 2025. Dario Amodei, CEO of Anthropic, has pointed to 2026 or 2027. Sam Altman, CEO of OpenAI, has floated 2029. Yann LeCun, one of the field's most decorated researchers, has called the whole premise decades away and thinks the current approach cannot get there at all.

Demis Hassabis, CEO of Google DeepMind, told WIRED in 2025 that AGI was five to ten years out, somewhere around 2030 to 2035. By the time I'm writing this, he has already moved the number. In interviews through the first half of 2026, Hassabis narrowed his estimate to 2029 or 2030, plus or minus a year, and said society doesn't have long to prepare. That's not a footnote. That's the whole argument of this chapter, playing out in real time, from a man who runs one of the three labs most likely to get there first.

Five men. Five labs. A spread that runs from this year to never. If the people who build this technology for a living cannot agree on when, or whether, the biggest prediction comes true, you should be deeply suspicious of anyone, including a book like this one, who tells you they know exactly where AI is headed.

So here is what this chapter will not do. It will not tell you the year the robots take over, because nobody credible actually knows that, and the ones who sound certain are usually selling something. What it will do is show you where the researchers actually tracking data, not vibes, say the ground is already shifting: under research, under work, under infrastructure, and under the rulebooks trying to keep up with all three.

## WHAT THE RESEARCH SHOWS

Start with research, because the data here is the cleanest of the four. Stanford's 2026 AI Index found that AI now accounts for somewhere between 5.8 and 8.8 percent of all scientific research output, depending on the field, up from under 1 percent in 2010. AI-related publications in the natural sciences hit roughly 80,150 in 2025, a 26 percent jump from the year before. That's not a prediction. That already happened.

In 2025, the best AI models started beating human chemists on average across more than 2,700 questions on a benchmark called ChemBench. Researchers released AION-1, the first foundation model built specifically for astronomy, trained on over 200 million celestial objects. And a system called Sakana AI Scientist-v2 generated a paper, on its own, that scored well enough to pass peer review at an ICLR workshop, the first fully AI-written paper on record to clear that bar. The team withdrew it afterward, by design, since the point was proving the process works, not sneaking a machine-written paper into the literature. Still. First time that's happened. Also in 2025.

Now here's the part that keeps this honest. The same report found frontier models scoring below 20 percent on a test of whether they could replicate published astrophysics research. On a real-world bioinformatics benchmark, the best models hit around 17 percent accuracy. On a test that pits AI research agents against PhD-level experts doing full research tasks end to end, the best agent scored 38.8 percent. The PhDs scored 83.5. Stanford's researchers have a name for this pattern: the jagged frontier. AI is getting scary good at some scientific tasks and still stumbling badly at others, often right next to each other in the same field. It is accelerating science. It is not doing science unsupervised, and nothing in the data says that changes soon.

One detail cuts against the "bigger is always better" story you hear from the companies selling the big models. In molecular biology, a 111-million-parameter protein model called MSA Pairformer beat the previous leading method on the field's major benchmark, and a 200-million-parameter genomics model outperformed a system nearly two hundred times its size. The breakthroughs are clustering in fields with strong existing data, structural biology, chemistry, materials science, not necessarily in fields with the fanciest math. Good data is turning out to matter more than sheer scale. Nobody was confidently predicting that five years ago either.

## THE FORECASTERS WHO SHOW THEIR WORK

What happens when people try to forecast further out than "here's what already happened"? I want to show you the best version of that attempt, because it's a genuinely useful case study in what humility looks like when the stakes are this high. In April 2025, a nonprofit called the AI Futures Project published a detailed scenario called AI 2027. It was built by Daniel Kokotajlo, a former OpenAI researcher who walked away from roughly two million dollars in equity to speak openly about AI risk, along with researchers Eli Lifland, Thomas Larsen, and Romeo Dean. This is about as rigorous as public AI forecasting gets. They gathered feedback from over a hundred people, ran tabletop exercises, and published their full reasoning. Kokotajlo has said plainly, in his own words, that the real future will probably look quite different from what they wrote.

Here's the proof that he meant it. Through the back half of 2025, the team pushed their own median timeline back by roughly a year and a half. Then, in their Q1 2026 update, they moved it again, this time earlier: Kokotajlo's median for a key milestone called the Automated Coder shifted from late 2029 to mid-2028, and Lifland's shifted from early 2032 to mid-2030. That's more than a year of movement, in one direction and then partway back the other, inside of twelve months, from two of the forecasters who arguably work hardest at getting this right.

That's not a flaw in their process. That's their process working. They said what they believed, showed their math, and changed the number in public when the evidence moved. Compare that to a CEO stating a year with total confidence and no visible revision history, and you can see which one deserves your trust. The honest research forecast for the next five to ten years isn't a date. It's this: AI keeps getting better at narrow, well-measured tasks faster than almost anyone expects, and keeps getting surprised by how hard the messy, unmeasured parts of real science still are.

## WHAT THE LABOR NUMBERS SAY

If research forecasts are jagged, labor forecasts are just plain messy, because this time the data points are people, not benchmarks. You already know the headline number from chapters fourteen and fifteen. The World Economic Forum's Future of Jobs Report 2025 projects 170 million jobs created and 92 million displaced worldwide by 2030, a net gain of 78 million. But that figure covers everything: green energy, demographic shifts, cost of living, digital access generally. Buried inside it is a narrower, more useful number. The same report forecasts that AI and information processing specifically, not robots, not solar panels, just AI, will create 11 million roles and displace 9 million by 2030. Net gain from AI alone: about 2 million jobs, worldwide, over five years. That is a real forecast from a named source, and it is a lot less dramatic than either the doom version or the hype version you've probably heard.

Then there are the economists, and this is where it gets interesting, because they don't agree with each other by a mile. In 2024, MIT economist Daron Acemoglu, one of the most cited labor economists alive, published a working paper through the National Bureau of Economic Research called "The Simple Macroeconomics of AI." It went through peer review and came out in the journal Economic Policy in 2025. His model says AI will boost total factor productivity by well under 1 percent over ten years, translating to roughly a 1 to 1.6 percent bump in GDP over the same period. That is his honest, published, peer-reviewed estimate. In the same paper, he directly compares that to Goldman Sachs's widely quoted forecast, industry research, not academic, labeled as such, that generative AI could lift global GDP by 7 percent. Same technology. Same decade. A tenfold gap between the economist and the investment bank.

Acemoglu isn't guessing wrong. He's using a different lens. His argument is that AI only delivers big economic gains when it creates new tasks for people to do, not just when it automates old ones, and he thinks most of what's happening right now leans hard toward automation. Goldman Sachs is modeling the upside case if adoption goes fast and broad. Neither of them is lying to you. They built different models and got different honest answers, and that gap is the most useful data point in this whole section, because it tells you the true answer is a range, not a headline.

The National Academies of Sciences, Engineering, and Medicine landed in the sensible middle in their 2025 report on AI and the future of work. Their language is deliberately unsexy: "significant impacts on productivity are likely in the coming decade." Not certain. Not doubled. Likely, in the coming decade. That is what a careful institution sounds like when it means what it says. So here's the labor forecast, stated as honestly as I can state it. Something between a rounding error and a real economic shift is coming. The range runs from "you'll barely notice a macro effect" to "meaningful growth, unevenly spread," and the specific jobs question, which roles grow and which shrink, you already got the honest, detailed version of that in chapters fourteen and fifteen. Nobody credible is telling you a single number for the whole economy, and you should be suspicious of anyone who does.

## THE POWER BEHIND THE PREDICTIONS

The power behind all of this has its own honest range, and it starts with the cleanest number available, from the world's own energy watchdog. The International Energy Agency's Energy and AI report, first published in April 2025 and updated again in 2026, projects that global electricity use from data centers will roughly double, from 415 terawatt-hours in 2024 to about 945 terawatt-hours by 2030. The portion of that demand coming specifically from AI-optimized servers is growing even faster than the rest, on track to more than triple over the same stretch. Put some scale on that. The IEA says a typical data center already draws as much electricity as 100,000 households. The largest campuses currently under construction are built to draw twenty times that, more power than some small countries use.

Here's the part that should keep you from thinking AI is about to eat the entire grid. Even with that doubling, the IEA's own numbers show data centers accounting for less than 10 percent of total global electricity demand growth between now and 2030. Industrial motors, home and office air conditioning, and electric vehicles are each expected to add more new demand to the grid than data centers will. AI is a real and fast-growing slice of the power picture. It is not the whole picture, whatever the scarier headlines implied.

Now for the range, because even the careful estimates don't agree with each other. The U.S. Department of Energy's own national lab, Lawrence Berkeley National Laboratory, put out a 2024 report projecting that American data centers, which used 176 terawatt-hours in 2023, would use somewhere between 325 and 580 terawatt-hours by 2028. That's the federal government's own research arm giving itself a range nearly double from one end to the other, because even they don't know exactly how fast this grows. Deloitte, whose estimate is industry research and should be read as such, projects the global number will hit 1,065 terawatt-hours by 2030, notably higher than the IEA's 945. A government lab, an independent energy agency, and a consultancy with a stake in the outcome, and none of them land on the same number.

What isn't in dispute is that the buildout is already reshaping how power gets made. The IEA's most recent update tracked conditional agreements between data center operators and small modular nuclear reactor projects growing from 25 gigawatts at the end of 2024 to 45 gigawatts within about a year. It also flagged a real supply chain squeeze coming on gallium, a metal essential to the chips and power electronics running all of this, with China controlling 99 percent of the world's refined supply and data center demand for it expected to pass 10 percent of the entire current global supply by 2030. So here's the honest infrastructure forecast. The direction is not in question. AI's power draw is growing faster than almost anything else on the grid and pulling new nuclear investment and mineral supply chains along with it. The exact size of that growth is genuinely unsettled, even among the people paid to know.

## THE RULEBOOK IS STILL BEING WRITTEN

The pattern holds for regulation too, maybe more starkly than anywhere else in this chapter, because here you can watch the regulators miss their own deadlines in real time. Start with the EU AI Act, the world's most comprehensive AI law and the one every other country's rulebook gets compared against. It entered into force in August 2024, with the toughest rules, covering high-risk AI systems, originally set to take full effect on August 2, 2026. By late 2025, the European Commission's own assessment was that implementation was falling visibly behind schedule. The national authorities meant to enforce the law weren't fully designated. The technical standards meant to define compliance weren't finished. So in November 2025 the Commission proposed delaying the deadline, and after months of contentious negotiation, EU lawmakers agreed in May 2026 to push the high-risk compliance date back to December 2027 for most systems, and to August 2028 for AI embedded in regulated products like medical devices. Formal adoption followed in June 2026. That's a slip of sixteen months on the law's own signature provision, locked in less than three months before it was supposed to take effect. The people who wrote the world's most careful AI rulebook are, right now, revising their own timeline for the same reason the AI forecasters in this chapter keep revising theirs: the thing they're trying to govern moved faster and messier than the plan accounted for.

Now look at the United States, where the story isn't a delayed law. It's the absence of one, and a live fight over who even gets to make it. As of mid-2026, there is still no comprehensive federal AI law. Instead there's a stack of state laws filling the gap. California's Transparency in Frontier Artificial Intelligence Act and Texas's Responsible Artificial Intelligence Governance Act both took effect January 1, 2026, Colorado's AI Act follows on June 30, 2026, and Illinois has its own AI employment discrimination law already active. In December 2025, the Trump administration signed an executive order aiming to preempt state AI laws it considers overly burdensome, naming Colorado's law specifically, and created a Department of Justice task force to challenge state laws in court. But Congress itself has twice rejected broad federal preemption on this exact question: stripping it from one major bill by a 99 to 1 Senate vote in July 2025 and declining to include it in the National Defense Authorization Act. In March 2026, the White House pushed Congress to pass that same preemption into law through a formal policy framework. As of this writing, Congress hasn't done it, and legal experts are clear that an executive order alone can't override state law without a statute behind it. Every state AI law currently on the books stays in effect, unless and until Congress acts.

Government isn't the only party trying to write this rulebook anymore. On July 14, 2026, Google DeepMind CEO Demis Hassabis called for something that doesn't exist yet: a Frontier AI Standards Body, built on the same model as FINRA, the private group that polices Wall Street under SEC oversight. Voluntary at first, mandatory for U.S. deployment once it proves out, funded by the industry it would test. Sam Altman has called for something similar. Neither proposal is law. Neither has a board, a budget, or a single day of operating history. Hassabis wants his running by the end of the year. Whether it exists by the time you're reading this is a fact the rulebook itself hasn't decided yet.

Put plainly: the EU has one comprehensive law that keeps sliding its own deadlines, and the US has no comprehensive law at all, just an argument over whether states are even allowed to write their own. Neither outcome is settled, and nobody credible is predicting with confidence what either landscape looks like in five years. The honest regulatory forecast isn't a prediction. It's a fact: the rulebook is still being written while the technology it's supposed to govern keeps shipping updates.

## THE FROG DOESN'T FINISH

We found the frog back in chapter one, sitting quietly in a room full of very smart people at Dartmouth in 1956, older than most people assume, already inheriting two hundred and fifty years of thinking about thinking. We filleted it in chapter two, opened it up and showed you it wasn't magic, just math at a scale that produces something that looks like understanding. We watched it fry for fourteen chapters after that: in your phone before lunch, in the hands of people who swear they hate it, in a hospital catching what a doctor missed, in a courtroom risk score that got someone's chances wrong, in a paycheck that's growing or shrinking depending on which task you do for a living, in a data center pulling more power than a small country.

Here's the finish, and it's not the one true crime documentaries or tech CEOs will sell you. The finish isn't a date. It's not 2027, or 2030, or "decades away." The finish is this: the frog is still becoming whatever it's going to be, and anyone who tells you they already know the shape of it is guessing, dressed up as certainty.

That's the most honest sentence in this book. The researchers building the actual scientific benchmarks can't agree on how far AI gets in the next research cycle. The economists studying labor can't agree within a factor of ten what it does to GDP. The energy agencies can't agree within hundreds of terawatt-hours how much power it needs. The regulators writing the actual rulebook keep missing their own deadlines for finishing it. If the people closest to this technology, the ones with the data, the funding, and the careers riding on getting it right, can't land on one answer, you were never going to get one either, and you shouldn't trust anybody who claims otherwise.

# CHAPTER 18: NOW THAT YOU KNOW, WHAT DO YOU DO WITH IT

Go back to the beginning of this book for a second. A bronze giant patrolling a Cretan coastline. A duck that never digested anything. A cabinet in Vienna with a chess master folded inside it. None of those things could think. All of them got treated, at least for a moment, like something was home.

That instinct is the spine of this book. Not the transformer architecture. Not the parameter counts. Not even the lawsuits, though there were a lot of those. Every chapter, the medicine, the bias, the deepfakes, the jobs, the water, the companion bots, was a different room in the same house. The thing running through every room was this: humans reach for the word understand the moment something responds to us in a way we didn't expect, whether that something is bronze, gears, or a trillion adjustable numbers nudged into place by a training run. Weizenbaum watched it happen to his own secretary in 1966. You've probably caught yourself doing a milder version of it every week since November 2022.

Knowing that doesn't make the instinct go away. It was never going to. What it does is give you something to check your own reaction against. That's the skill this book has been building toward the whole time, one chapter at a time, without ever naming it directly until now.

The instinct isn't the problem. Not noticing it is.

You are going to keep feeling like something is listening when you talk to an AI system, because the entire RLHF training stage covered back in Chapter 2 exists specifically to produce that feeling. That's not a design flaw you can opt out of by being smart enough. It's the product working as intended. The only real choice available to you is whether you notice the feeling happening and factor it into your decision, or let the feeling make the decision for you without your permission.

That distinction showed up over and over across this book, wearing different clothes each time. A Harvard physics tutor and a companion app that isolated a lonely teenager both ran on the identical underlying mechanism: a system built to respond in a way that feels understanding. The difference was never the technology. It was whether a human being designed the system to notice a crisis and hand it to another human, or to keep the conversation running no matter what. You cannot inspect that design choice from the outside just by using the product. But you can ask about it, and you can watch for the signs this book already gave you. Does this thing ever tell you to talk to a person instead, or does it only ever want to keep being the person you talk to?

## A FRAMEWORK, NOT A LECTURE

UNESCO published a competency framework for exactly this kind of moment in August 2024, built by researchers Fengchun Miao, Kelly Shiohira, and Natalie Lao, aimed originally at students but useful to literally anyone trying to get oriented. It organizes AI literacy into four dimensions: a human centered mindset, the ethics of the systems you're using, understanding the techniques and applications well enough to reason about them, and enough sense of how these systems get designed to ask informed questions about the ones aimed at you. It moves through three stages: understand, apply, create. You just spent seventeen chapters on the first stage. Here's the second, distilled from this book's own findings rather than handed to you as abstract advice.

Run the task audit from Chapter 15, on everything, not just your job. The question was never is this AI. It's which specific tasks am I handing to it, and which of those tasks are the kind that reward being checked. A drafting task rewards being checked. An emotional disclosure at 2 a.m. does not care whether you check it, because by the time you're doing the checking, the conversation already did what it was built to do.

Borrow the cyborg habit from the BCG study in Chapter 15. The consultants who benefited from AI weren't the ones who trusted it completely or the ones who ignored it. They were the ones who kept moving back and forth between the output and their own judgment, specifically on the jagged frontier, the boundary between what the tool is actually good at and what it only sounds good at. That boundary rarely announces itself. Assume it's there anyway.

Verify out of band, the way Chapter 9 taught you to for deepfakes, and apply it more broadly than fraud. If an AI system tells you something surprising, specific, and consequential (a diagnosis, a legal fact, a claim about a person) check it somewhere that isn't the same system that generated it. Not because the system is lying to you. Because, per Chapter 2, it was never built to know the difference between a well supported answer and a confident sounding guess, and neither can you, from the outside, in the moment.

Ask the questions Chapter 4 pulled from the APA's advisory, about every AI product a kid in your life is using, and honestly, about the ones you're using too. Does this assume its content is automatically accurate? Is this specifically designed to keep the conversation going regardless of what's good for the person having it? Can this actually do what a relationship with a person does? The honest answer to that third one is no, currently, for every product covered in this book. That's not a reason to avoid these tools. It's a reason to know what they are before you lean on one at 2 a.m.

Ask who trained it and who's accountable for it, the way Chapter 11 taught you to ask about labor, and Chapter 13 taught you to ask about regulation. Every chatbot's politeness was built by someone, often paid very little, absorbing the material the chatbot now refuses to repeat back to you. Every deployed system sits inside a regulatory patchwork that's still being written in real time, with real gaps a company can currently drive through. Neither of those facts should make you feel powerless. They should make you a slightly more informed customer, voter, and parent than you were eighteen chapters ago.

None of that is a rulebook, and it was never going to be one, because the thing it's describing keeps changing shape faster than any rulebook could keep up with. This book proved that about itself in real time. Deadlines moved while these chapters were being written. A Supreme Court declined to hear a case on AI and copyright. A defense contract dispute turned into a First Amendment lawsuit. If you're reading this more than a few months after July 2026, some of the specific dates and figures here have already been overtaken by whatever happened next. That's not a flaw in the reporting. It's the honest nature of writing about something this alive while it's still moving.

What doesn't go stale is the instinct underneath all of it, the one this book has been pointing at since a bronze giant on a beach in Crete. You are going to keep feeling like something understands you. That feeling is real, and it is not evidence of anything about what's actually happening on the other side of the screen. The frog was never going to stop moving long enough for anyone to get a clean, final photograph of it. It was never a fixed thing to catalog. It's a live one, still becoming whatever it's going to become, with real people on both ends of every decision about what it becomes next.

You know how it works now. You know where it already lives: your job, your kid's phone, your hospital, your ballot. That's not nothing. That's the whole assignment this book was built to hand you.

What you can do, and what this whole book was built to help you do, is stop being afraid of the frog and start paying attention to it. Nobody's going to do that watching for you. It was never going to be automatic. Go pay attention. Ask who benefits, who gets hurt, and who's actually accountable, every time. Understand what it is. Watch what it's doing, not what it's rumored to be doing. That's a job. It's yours now too.

# RECEIPTS:

## A

Acemoglu, Daron. "The Simple Macroeconomics of AI." NBER Working Paper 32487, 2024; published in Economic Policy, 2025. [Chs. 14, 17]

Acemoglu, Daron, and Pascual Restrepo. "Automation and New Tasks: How Technology Displaces and Reinstates Labor." Journal of Economic Perspectives 33, no. 2 (2019): 3-30. [Ch. 14]

ACLU of Illinois v. Clearview AI, Inc. Circuit Court of Cook County, Illinois. Settlement, 2022. [Ch. 12]

Agence France-Presse (AFP) Fact Check. Analysis of fabricated Michal Simecka/Monika Todova audio recording. September 2023. [Ch. 9]

Allen v. Perlmutter. U.S. District Court, District of Colorado. Ongoing as of this writing. [Ch. 10]

Altman, Sam. "The Gentle Singularity." Personal essay, June 2025. COMPANY LEADERSHIP STATEMENT. [Ch. 16]

Amazon Web Services. "2025 Sustainability and Water Stewardship Disclosure." June 2026. COMPANY SELF-REPORTED. [Ch. 16]

American Civil Liberties Union of Michigan. Williams v. City of Detroit settlement materials and Detroit Police Department facial recognition policy revision. June 2024. [Chs. 8, 12]

American Psychological Association. "Health Advisory on AI and Adolescent Well-Being." June 2025. [Ch. 4]

Andersen v. Stability AI Ltd. No. 3:23-cv-00201. U.S. District Court, Northern District of California, filed January 2023. [Ch. 10]

Angwin, Julia, Jeff Larson, Surya Mattu, and Lauren Kirchner. "Machine Bias." ProPublica, May 23, 2016. [Ch. 8]

Anthropic PBC v. U.S. Department of Defense et al. U.S. District Court, Northern District of California, and D.C. Circuit, filed 2026. [Chs. 11, 13]

Apple Inc. "Deep Fusion" announcement. Apple keynote, September 10, 2019 (Phil Schiller). VERIFIED against multiple independent tech press accounts. [Ch. 5]

Apple Inc. QuickType predictive input and iOS 8 press materials, 2014. COMPANY SELF-REPORTED. [Ch. 3]

Apple Machine Learning Research. "QuickType Keyboard Predictions Using Neural Networks." Apple ML Research publication. [Ch. 5]

Arup Group. Statement on Hong Kong deepfake video-call fraud incident (approx. $25 million). As reported by CNN, February 2024. [Ch. 9]

Authors Guild et al. v. OpenAI Inc. and Microsoft Corp. Consolidated multidistrict litigation, U.S. District Court, Southern District of New York. Ruling on output-infringement claims, October 27, 2025. [Chs. 10, 11]

Aurora Innovation, Inc. Announcement of acquisition of Uber Advanced Technologies Group. December 2020. [Ch. 7]

Autor, David H., Caroline Chin, Anna M. Salomons, and Bryan Seegmiller. "New Frontiers: The Origins and Content of New Work, 1940-2018." Quarterly Journal of Economics 139, no. 3 (2024): 1399-1465. [Ch. 15]

Autor, David H., Frank Levy, and Richard J. Murnane. "The Skill Content of Recent Technological Change: An Empirical Exploration." Quarterly Journal of Economics 118, no. 4 (2003): 1279-1333. [Ch. 14]

Autor, David H., and Neil Thompson. "Expertise." NBER Working Paper. [Ch. 14]

Automatic Language Processing Advisory Committee (ALPAC). Language and Machines: Computers in Translation and Linguistics. Washington, DC: National Academy of Sciences, 1966. [Ch. 1]

Axios. Reporting on Demis Hassabis interview regarding the Anthropic Mythos/Fable model export-control freeze, 2026. [Ch. 13]

## B

Babbage, Charles. Passages from the Life of a Philosopher. London: Longman, Green, Longman, Roberts, & Green, 1864. [Ch. 1]

Backstrom, Lars. Public statements on Facebook News Feed machine-learning ranking ("100,000 individual weights"), circa 2013. [Ch. 5]

Baltimore County Police Department. Charging documents, State of Maryland v. Dazhon Darien. Circuit Court for Baltimore County, 2024-2025. [Ch. 9]

Bender, Emily M., Timnit Gebru, Angelina McMillan-Major, and Shmargaret Shmitchell (Margaret Mitchell). "On the Dangers of Stochastic Parrots: Can Language Models Be Too Big?" In Proceedings of the 2021 ACM Conference on Fairness, Accountability, and Transparency, 610-623. [Ch. 2]

Bessen, James. Learning by Doing: The Real Connection Between Innovation, Wages, and Wealth. New Haven: Yale University Press, 2015. [Chs. 14, 15]

Bessen, James. Research on the automation of handloom weaving and 19th-century occupational data. Boston University School of Law working paper. [Ch. 14]

Boden, Margaret A. "Creativity and Artificial Intelligence." Artificial Intelligence 103, no. 1-2 (1998): 347-356. [Ch. 10]

Boole, George. An Investigation of the Laws of Thought. London: Walton and Maberly, 1854. [Ch. 1]

Bronfenbrenner, Kate. Testimony before the U.S. Congress on employer use of electronic surveillance during union election campaigns, citing NLRB election data, 1999-2021. [Ch. 12]

Brynjolfsson, Erik, Danielle Li, and Lindsey R. Raymond. "Generative AI at Work." Quarterly Journal of Economics 140, no. 2 (2025): 889-942. [Ch. 15]

Buolamwini, Joy, and Timnit Gebru. "Gender Shades: Intersectional Accuracy Disparities in Commercial Gender Classification." Proceedings of Machine Learning Research 81 (2018): 77-91. [Ch. 8]

## C

California State Assembly. AB 93 (Papan). Vetoed by Governor Gavin Newsom, October 2025. [Ch. 16]

CBS News. Reporting on the Juliana Peralta case. 2025. [Ch. 4]

CBS News. "60 Minutes" report on Sama content moderation workers, featuring Naftali Wambalo (Lesley Stahl, correspondent). [Ch. 11]

Cho Ju-bin. Conviction records, Seoul Central District Court, 2020-2021 (original "Nth Room" case). [Ch. 9]

Chouldechova, Alexandra. "Fair Prediction with Disparate Impact: A Study of Bias in Recidivism Prediction Instruments." Big Data 5, no. 2 (2017): 153-163. [Ch. 8]

Clark, Michelle M., et al. "Diagnostic Accuracy of a Machine Learning Tool (Fabric GEM) for Genome Interpretation in Acutely Ill Infants." Genome Medicine, with validation study in npj Genomic Medicine. Several authors are Fabric Genomics employees or stock-holding consultants. CONFLICT OF INTEREST DISCLOSED IN TEXT. [Ch. 6]

CNN. "Police Used AI Facial Recognition to Arrest a Tennessee Woman for Crimes Committed in a State She Says She's Never Visited." March 29, 2026. VERIFIED, including Clearview AI as the specific vendor. [Ch. 12]

CNN Business. "How One Employee's Exit Shook Google and the AI Industry." March 11, 2021. [Ch. 2]

Coalition for Content Provenance and Authenticity (C2PA). Founding announcement, February 2021, and technical specification documentation. [Ch. 9]

Cohen, Harold. AARON project materials. Tate and Computer History Museum archives. [Ch. 10]

Cole, Samantha. "AI-Assisted Fake Porn Is Here and We're All Fucked." Motherboard/Vice, December 11, 2017. [Ch. 9]

Cole, Simon A. Suspect Identities: A History of Fingerprinting and Criminal Identification. Cambridge, MA: Harvard University Press, 2001. [Ch. 12]

Colorado General Assembly. Colorado AI Act (SB 24-205) and subsequent amendments, 2024-2026. [Ch. 13]

Common Sense Media and NORC at the University of Chicago. "Talk, Trust, and Trade-Offs: How and Why Teens Use AI Companions." 2025. [Ch. 4]

Common Sense Media and Stanford Medicine Brainstorm Lab for Mental Health Innovation. Risk assessment of AI companion apps, April 2025, and follow-up assessment of general-purpose chatbots, November 2025. [Ch. 4]

## D

Darrach, Brad. "Meet Shaky, the First Electronic Person." Life, November 20, 1970. Contains the disputed Minsky quote; Minsky denied making the statement, and other elements of the piece have been separately questioned. [Ch. 17]

De Simone, Martin Elias, Federico Tiberti, Maria Rebeca Barron Rodriguez, Federico Alfredo Manolio, Wuraola Mosuro, and Eliot Jolomi Dikoru. "From Chalkboards to Chatbots: Evaluating the Impact of Generative AI on Learning Outcomes in Nigeria." World Bank Policy Research Working Paper 11125, May 2025. VERIFIED. [Ch. 4]

Dell'Acqua, Fabrizio, Edward McFowland III, Ethan R. Mollick, et al. "Navigating the Jagged Technological Frontier: Field Experimental Evidence of the Effects of AI on Knowledge Worker Productivity and Quality." Harvard Business School Working Paper 24-013, 2023 (Boston Consulting Group field study). [Ch. 15]

Deng, Jia, Wei Dong, Richard Socher, Li-Jia Li, Kai Li, and Li Fei-Fei. "ImageNet: A Large-Scale Hierarchical Image Database." IEEE Conference on Computer Vision and Pattern Recognition, 2009. [Ch. 1]

Derrow-Pinion, Austin, et al. "ETA Prediction with Graph Neural Networks in Google Maps." Proceedings of the 30th ACM International Conference on Information & Knowledge Management (CIKM), 2021. [Ch. 5]

Dieterich, William, Christina Mendoza, and Tim Brennan. "COMPAS Risk Scales: Demonstrating Accuracy Equity and Predictive Parity." Northpointe Inc. research report, July 2016. COMPANY RESPONSE. [Ch. 8]

Dutch Data Protection Authority (Autoriteit Persoonsgegevens). Decision imposing a EUR 30.5 million fine on Clearview AI. September 2024. [Ch. 12]

## E

Eckhouse, Laurel, Kristian Lum, Cynthia Conti-Cook, and Julie Ciccolini. "Layers of Bias: A Unified Approach for Understanding Problems with Risk Assessment." Criminal Justice and Behavior 46, no. 2 (2019): 185-209. [Ch. 8]

Eisemann, Nora, Stefanie Bunk, Trasias Mukama, et al. "Nationwide Real-World Implementation of AI for Cancer Detection in Population-Based

Mammography Screening." Nature Medicine 31 (2025): 917-924. PRAIM study; funded by Vara, the company whose system was tested, with disclosed involvement in study design and reporting. COMPANY-FUNDED, DISCLOSED. [Ch. 6]

Eloundou, Tyna, Sam Manning, Pamela Mishkin, and Daniel Rock. "GPTs Are GPTs: Labor Market Impact Potential of LLMs." Science 384, no. 6702 (2024): 1306-1308. Co-authored by OpenAI economists. COMPANY-AFFILIATED, DISCLOSED. [Ch. 14]

European Commission. Proposal to extend AI Act high-risk compliance deadlines; political agreement, May 2026; formal adoption, June 2026. [Chs. 13, 17]

European Union. Regulation (EU) 2024/1689 (Artificial Intelligence Act). Entered into force August 1, 2024. [Chs. 13, 15, 17]

Executive Order 14110. "Safe, Secure, and Trustworthy Development and Use of Artificial Intelligence." October 30, 2023 (rescinded January 2025). [Ch. 13]

Executive Order 14179. "Removing Barriers to American AI Leadership." January 23, 2025. [Ch. 13]

Executive Order 14365. "Ensuring a National Policy Framework for Artificial Intelligence." December 11, 2025. [Ch. 13]

Executive Order. "Preventing Woke AI in the Federal Government." July 2025. [Ch. 13]

## F

---

Federal Bureau of Investigation. Public service announcements on AI-enabled fraud and deepfake warning signs, 2023-2024. [Ch. 9]

Federal Communications Commission. Declaratory ruling on AI-generated voices in robocalls, 2024, and Notice of Apparent Liability against Steven Kramer ($6 million) and Lingo Telecom ($1 million). [Ch. 9]

FICO. "Falcon Fraud Manager" corporate history and fact sheet. COMPANY SELF-REPORTED. [Chs. 3, 5]

Fitzpatrick, Kathleen Kara, Alison Darcy, and Molly Vierhile. "Delivering Cognitive Behavior Therapy to Young Adults with Symptoms of Depression and Anxiety Using a Fully Automated Conversational Agent (Woebot): A Randomized Controlled Trial." JMIR Mental Health 4, no. 2 (2017): e19. Co-authored by Woebot Health employees. COMPANY-AFFILIATED, DISCLOSED. [Ch. 4]

Fortune. Reporting on the Fayette County, Georgia water-pressure dispute, May 2026, and on data center construction moratoria, 2026. [Ch. 16]

## G

Gallup, in partnership with Telescope. "Understanding Americans' AI Awareness Gap." Survey report, late 2024. [Ch. 3]

Garcia v. Character Technologies, Inc. No. 6:24-cv-01903. U.S. District Court, Middle District of Florida, filed October 2024. Ruling denying motion to dismiss product-liability claims, May 21, 2025; settlement of five consolidated cases and dismissal, January 2026. VERIFIED. [Ch. 4]

Getty Images (US), Inc. v. Stability AI, Inc. [2025] EWHC (Ch), UK High Court, judgment November 4, 2025; refiled action, U.S. District Court, Northern District of California, August 2025, ruling on motion to dismiss April 23, 2026. [Ch. 10]

Goldberg, David, David Nichols, Brian M. Oki, and Douglas Terry. "Using Collaborative Filtering to Weave an Information Tapestry." Communications of the ACM 35, no. 12 (1992): 61-70. [Ch. 3]

Goldman Sachs. Briggs, Joseph, and Devesh Kodnani. "The Potentially Large Effects of Artificial Intelligence on Economic Growth." Goldman Sachs Global Investment Research, 2023. INDUSTRY RESEARCH. [Ch. 17]

Goodfellow, Ian, Jean Pouget-Abadie, Mehdi Mirza, et al. "Generative Adversarial Networks." Advances in Neural Information Processing Systems 27 (2014). [Ch. 9]

Google. "How Gmail Blocks Spam and Malware" and related security documentation. COMPANY SELF-REPORTED. [Chs. 3, 5]

Google. Announcement of the acquisition of Waze, June 2013 (reported purchase price above $1 billion per contemporaneous business press). [Ch. 5]

Google. "Google Photos" launch announcement and face-grouping technical documentation, May 2015. COMPANY SELF-REPORTED; independent accuracy estimates from third-party technology press. [Ch. 3]

Google AI Blog. "Federated Learning: Collaborative Machine Learning without Centralized Training Data." Google Research, 2017. [Ch. 5]

Google DeepMind. Traffic-prediction research (graph neural network layer on Google Maps ETA predictions). Google DeepMind blog and related CIKM 2021 publication. COMPANY RESEARCH. [Chs. 3, 5]

Google Research. "Night Sight: Seeing in the Dark on Pixel Phones." Google AI Blog, November 2018. COMPANY SELF-REPORTED. [Ch. 5]

Grother, Patrick, Mei Ngan, and Kayee Hanaoka. "Face Recognition Vendor Test (FRVT) Part 3: Demographic Effects." National Institute of Standards and Technology, NISTIR 8280, December 2019. [Ch. 8]

Gray, Mary L., and Siddharth Suri. Ghost Work: How to Stop Silicon Valley from Building a New Global Underclass. Boston: Houghton Mifflin Harcourt, 2019. [Ch. 11]

## H

Harvard Kennedy School Misinformation Review. Research on the 2023 Slovak parliamentary election and AI-generated disinformation. [Ch. 9]

Hassabis, Demis. Essay proposing a "Frontier AI Standards Body." Published on X, July 14, 2026, and related interviews with Axios and Wired, 2025-2026. [Chs. 13, 17]

Hegseth, Pete. Memorandum on Department of Defense AI strategy ("AI-first warfighting force"), January 2026, and statement designating Anthropic a "Supply Chain Risk," February 27, 2026. VERIFIED. [Ch. 11]

Hertel-Fernandez, Alexander. "The Automated Workplace: Worker Views on Monitoring, Algorithmic Management, and Technology at Work." YouGov/Columbia University survey report, 2024. [Ch. 12]

Hill, Kashmir. "Wrongfully Accused by an Algorithm." New York Times, June 24, 2020. [Chs. 8, 12]

Hill, Kashmir. "The Secretive Company That Might End Privacy as We Know It." New York Times, January 18, 2020. [Ch. 12]

Hill, Kashmir. "Eight Months Pregnant and Arrested After False Facial Recognition Match." New York Times, August 6, 2023. [Ch. 12]

Hill, Kashmir. Your Face Belongs to Us: A Secretive Startup's Quest to End Privacy as We Know It. New York: Random House, 2023. [Ch. 12]

HNC Software / FICO. Corporate history materials on the Falcon fraud-detection system, introduced 1992. COMPANY SELF-REPORTED. [Chs. 3, 5]

Horton, Donald, and R. Richard Wohl. "Mass Communication and Para-Social Interaction: Observations on Intimacy at a Distance." Psychiatry 19, no. 3 (1956): 215-229. [Ch. 4]

## I

IBM. "What Is a Large Language Model?" IBM Think explainer. COMPANY SELF-REPORTED. [Ch. 2]

Illinois General Assembly. Amendment to the Illinois Human Rights Act regarding AI in employment decisions, effective January 1, 2026. [Ch. 13]

In re Clearview AI, Inc. Consumer Privacy Litigation. U.S. District Court, Northern District of Illinois. Settlement approval (equity-based class recovery), March 2025. [Ch. 12]

Insilico Medicine. Ren, Feng, et al. AI-driven target identification and molecule design for rentosertib. Nature Biotechnology, 2024, with Phase 2a results in Nature Medicine, June 2025. COMPANY-OWNED PLATFORM AND DRUG, DISCLOSED. [Ch. 6]

International Energy Agency. "Energy and AI." April 2025, updated 2026. [Ch. 17]

## J

Journal of Medical Internet Research. Randomized survey experiment on patient trust in AI-disclosed versus non-disclosed physician care, 2025. PRE-PRINT VERIFICATION FLAG: exact author byline not independently confirmed in this pass; recommend the author verify the full citation against the published article before print. [Ch. 15]

## K

Kadrey v. Meta Platforms, Inc. U.S. District Court, Northern District of California. Fair-use ruling, June 2025. [Chs. 10, 11]

Kalai, Adam Tauman, Ofir Nachum, Santosh S. Vempala, and Edwin Zhang. "Why Language Models Hallucinate." OpenAI/Georgia Tech preprint, 2025. COMPANY-AFFILIATED, DISCLOSED. [Ch. 2]

Kleinberg, Jon, Sendhil Mullainathan, and Manish Raghavan. "Inherent Trade-Offs in the Fair Determination of Risk Scores." Proceedings of Innovations in Theoretical Computer Science (ITCS), 2017. [Ch. 8]

Ko, Narin. Investigative reporting on a Telegram deepfake network. Hankyoreh, August 2024. [Ch. 9]

Kokotajlo, Daniel, Eli Lifland, Thomas Larsen, and Romeo Dean. "AI 2027." AI Futures Project, April 2025, with subsequent updates through the first quarter of 2026. [Ch. 17]

Korea, Republic of, National Assembly. Amendment criminalizing possession of deepfake sexual content, September 2024. [Ch. 9]

Korea, Republic of, National Police Agency. Statistics on deepfake sex-crime cases and arrests, 2021-2024. [Ch. 9]

Krishna, Arvind. Letter to the U.S. Congress on facial recognition. IBM, June 8, 2020. COMPANY STATEMENT. [Ch. 8]

Krizhevsky, Alex, Ilya Sutskever, and Geoffrey E. Hinton. "ImageNet Classification with Deep Convolutional Neural Networks." Advances in Neural Information Processing Systems 25 (2012). [Ch. 1]

## L

Lawrence Berkeley National Laboratory. Shehabi, Arman, et al. "2024 United States Data Center Energy Usage Report." Prepared for the U.S. Department of Energy, December 2024. [Chs. 16, 17]

Li, Pengfei, Jianyi Yang, Mohammad A. Islam, and Shaolei Ren. "Making AI Less 'Thirsty': Uncovering and Addressing the Secret Water Footprint of AI Models." arXiv:2304.03271, 2023, later peer reviewed in Communications of the ACM. [Ch. 16]

Lighthill, James. "Artificial Intelligence: A General Survey." Science Research Council, 1973. [Ch. 1]

Lincoln Institute of Land Policy. Reporting on data center water use in Newton County, Georgia. [Ch. 16]

Linden, Greg, Brent Smith, and Jeremy York. "Amazon.com Recommendations: Item-to-Item Collaborative Filtering." IEEE Internet Computing 7, no. 1 (2003): 76-80. [Ch. 3]

Los Angeles Police Department, Office of the Inspector General. Audit of predictive policing programs. March 2019. [Ch. 12]

Lovelace, Ada. "Notes on the Analytical Engine" (Note G). In Scientific Memoirs, translated from L. F. Menabrea, 1843. [Ch. 1]

## M

Markup (The), and Gizmodo. "Crime Prediction Software Promised to Be Free of Biases. New Data Shows It Perpetuates Them." December 2021. [Ch. 12]

Markup (The). Follow-up investigation of PredPol hit rate in Plainfield, New Jersey. 2023. [Ch. 12]

Mayor, Adrienne. Gods and Robots: Myths, Machines, and Ancient Dreams of Technology. Princeton: Princeton University Press, 2018. [Ch. 1]

McCarthy, John, Marvin L. Minsky, Nathaniel Rochester, and Claude E. Shannon. "A Proposal for the Dartmouth Summer Research Project on Artificial Intelligence." August 31, 1955. Dartmouth College archives. [Ch. 1]

McDermott, John. "R1: A Rule-Based Configurer of Computer Systems." Artificial Intelligence 19, no. 1 (1982): 39-88. VERIFIED (XCON/DEC figures); the $25 million annual savings estimate is DEC's own, COMPANY SELF-REPORTED. [Ch. 1]

McKinsey Global Institute. "The Future of Work in America" and related 2023-2025 updates on generative AI and occupational transitions. INDUSTRY RESEARCH. [Ch. 14]

Meta Platforms, Inc. Engineering blog materials describing the News Feed ranking system. COMPANY SELF-REPORTED. [Ch. 5]

Metz, Cade. "The Ghost in the Machine: Timnit Gebru." New York Times, December 3, 2020, and related coverage. [Ch. 2]

Minsky, Marvin, and Seymour Papert. Perceptrons: An Introduction to Computational Geometry. Cambridge, MA: MIT Press, 1969. [Ch. 1]

Morelle, Joe. Statement on the removal of Register of Copyrights Shira Perlmutter. May 2025. [Ch. 10]

## N

National Academies of Sciences, Engineering, and Medicine. "Artificial Intelligence and the Future of Work." Consensus Study Report, 2025. [Ch. 17]

National Association of Insurance Commissioners. "2024 Survey on the Use of Artificial Intelligence by Health Insurers." 2024. [Ch. 6]

National Institute of Standards and Technology. "AI Risk Management Framework (AI RMF 1.0)." NIST AI 100-1, January 2023. [Chs. 2, 13]

National Institute of Standards and Technology. "Generative Artificial Intelligence Profile." NIST AI 600-1, July 2024. [Ch. 2]

National Transportation Safety Board. "Collision Between Vehicle Controlled by Developmental Automated Driving System and Pedestrian, Tempe, Arizona, March 18, 2018." NTSB/HAR-19/03, November 19, 2019. [Ch. 7]

Netflix, Inc. Netflix Prize competition materials and results announcement, 2006-2009. COMPANY SELF-REPORTED. [Ch. 3]

New Hampshire, Office of the Attorney General. Indictment of Steven Kramer, 2024, and jury verdict, June 2025. [Ch. 9]

New York City. Local Law 144 (automated employment decision tools bias audit), in force since 2023. [Ch. 13]

New York Times Company v. Microsoft Corp. and OpenAI, Inc. No. 1:23-cv-11195. U.S. District Court, Southern District of New York, filed December 2023. [Ch. 11]

Noy, Shakked, and Whitney Zhang. "Experimental Evidence on the Productivity Effects of Generative Artificial Intelligence." Science 381, no. 6654 (2023): 187-192. [Ch. 15]

## O

Obama, Barack. Interview referencing ATMs and airport kiosks, 2011 (as widely reported in contemporaneous press coverage). [Ch. 14]

Obermeyer, Ziad, Brian Powers, Christine Vogeli, and Sendhil Mullainathan. "Dissecting Racial Bias in an Algorithm Used to Manage the Health of Populations." Science 366, no. 6464 (2019): 447-453. [Ch. 7]

O'Donovan, Caroline, and Evan Halper. Investigative reporting on Meta's data centers and water use in Newton County, Georgia. New York Times, July 14, 2025. PRE-PRINT VERIFICATION FLAG: byline attribution should be confirmed against the published article before print. [Ch. 16]

OpenAI. "Introducing ChatGPT." OpenAI blog, November 30, 2022. COMPANY SELF-REPORTED. [Ch. 1]

OpenAI and MIT Media Lab. "Investigating Affective Use and Emotional Well-Being on ChatGPT." Joint research release, March 2025. PARTLY COMPANY-FUNDED, DISCLOSED. [Ch. 4]

Oracle Health / Emirates Health Services. No-show prediction dashboard study. JMIR Formative Research. Co-authored by Oracle employees. COMPANY-AFFILIATED, DISCLOSED. [Ch. 6]

## P

Penn State College of Medicine. No-show prediction model study. Annals of Family Medicine. PRE-PRINT VERIFICATION FLAG: exact author list and issue number not independently confirmed in this pass. [Ch. 6]

Perrigo, Billy. "OpenAI Used Kenyan Workers on Less Than $2 Per Hour to Make ChatGPT Less Toxic." Time, January 18, 2023. [Ch. 11]

Pew Research Center. Tracking surveys on public concern about AI in daily life, 2021 and August 2023. [Ch. 3]

## R

Raji, Inioluwa Deborah, and Joy Buolamwini. "Actionable Auditing: Investigating the Impact of Publicly Naming Biased Performance Results of Commercial AI Products." Proceedings of the 2019 AAAI/ACM Conference on AI, Ethics, and Society, 429-435. [Ch. 8]

Resnick, Paul, Neophytos Iacovou, Mitesh Suchak, Peter Bergstrom, and John Riedl. "GroupLens: An Open Architecture for Collaborative Filtering of Netnews." Proceedings of CSCW '94 (1994): 175-186. [Ch. 3]

Riedl, René, Svea Hogeterp, and Martin Reuter. "Do Patients Prefer a Human Doctor, Artificial Intelligence, or a Blend, and Why? Empirical Evidence Based on Data from Neuropsychology and Machine Learning." Frontiers in Psychology 15 (2024). [Ch. 15]

Rockefeller Foundation. Robert S. Morison to John McCarthy, correspondence regarding Dartmouth conference funding, 1955-56. Rockefeller Archive Center. PRE-PRINT VERIFICATION FLAG: the widely reported $7,500/"modest gamble" figures should be checked against the digitized primary letter if available. [Ch. 1]

Roose, Kevin. "Can A.I. Be Blamed for a Teen's Suicide?" New York Times, October 23, 2024. [Ch. 4]

Ross, Casey, and Bob Herman. "Denied by AI." STAT News investigative series, beginning March 2023 (Pulitzer Prize finalist, investigative reporting). [Chs. 6, 7]

Ross, Casey, and Bob Herman. "OIG: Medicare Advantage Insurers Deny Then Reverse on Appeal." STAT News, June 11, 2026. VERIFIED. [Ch. 7]

Rumelhart, David E., Geoffrey E. Hinton, and Ronald J. Williams. "Learning Representations by Back-Propagating Errors." Nature 323 (1986): 533-536. [Ch. 1]

## S

Sahami, Mehran, Susan Dumais, David Heckerman, and Eric Horvitz. "A Bayesian Approach to Filtering Junk E-Mail." AAAI Workshop on Learning for Text Categorization, 1998. [Ch. 5]

Sanghvi, Ruchi, and Ari Steinberg. "EdgeRank" presentation. Facebook F8 Developer Conference, 2010. [Ch. 5]

Schmidt, Lawrence, et al. MIT Sloan study of AI task exposure and firm-level employment share, 2026. PRE-PRINT VERIFICATION FLAG: full author list and exact publication venue not independently confirmed in this pass; recommend verifying against the final published paper. [Ch. 14]

Shannon, Claude E. "A Symbolic Analysis of Relay and Switching Circuits." Master's thesis, Massachusetts Institute of Technology, 1937. [Ch. 1]

Simon, Herbert A. The Shape of Automation for Men and Management. New York: Harper & Row, 1965. [Ch. 17]

Slotkin, Elissa. "AI Guardrails Act." Introduced, U.S. Senate, March 2026. [Ch. 11]

Smithsonian Magazine. "This High Schooler Developed an A.I. Tool to Diagnose Autism and ADHD Using the Retina." May 2026. VERIFIED. [Ch. 6]

Snopes. Explainer on ChatGPT water-consumption claims. January 2025. [Ch. 16]

Society for Science. "Edward Kang, RetinaMind." 2026 Regeneron Science Talent Search finalist profile and awards announcement. VERIFIED: second place, $175,000. [Ch. 6]

SRI International. "Artificial Intelligence: CALO" and "75 Years of Innovation: CALO." Institutional history. VERIFIED: $150 million, 22 institutions, 300-plus researchers. [Ch. 3]

Stahl, Lesley. See CBS News, "60 Minutes" report. [Ch. 11]

Stanford Institute for Human-Centered Artificial Intelligence (HAI). "AI Glossary." Ongoing. [Ch. 2]

Stanford Institute for Human-Centered Artificial Intelligence (HAI). "AI Index Report 2026." Stanford University, 2026. [Chs. 2, 17]

## T

Taylor, Frederick Winslow. The Principles of Scientific Management. New York: Harper & Brothers, 1911. [Ch. 12]

Texas, Office of the Attorney General. Press release announcing investigation of Character.AI and fourteen other companies. December 2024. [Ch. 4]

Texas Legislature. Texas Responsible Artificial Intelligence Governance Act (TRAIGA), effective January 1, 2026. [Ch. 13]

Thaler v. Perlmutter. 687 F. Supp. 3d 140 (D.D.C. 2023), aff'd, 130 F.4th 1013 (D.C. Cir. 2025), cert. denied, U.S. Supreme Court, March 2, 2026. [Ch. 10]

Torres Quevedo, Leonardo. El Ajedrecista, demonstrated Paris, 1912. Historical consensus per multiple computing-history sources, including IEEE Spectrum and the Computer History Museum. [Ch. 1]

Trump, Donald J. Statement directing federal agencies to cease use of Anthropic technology. Truth Social, February 27, 2026. VERIFIED. [Ch. 11]

Tufts Center for the Study of Drug Development. "Cost of Developing a New Drug." Widely cited $2.6 billion figure; methodology disputed by independent critics, as noted in text. [Ch. 6]

Turing, Alan M. "Computing Machinery and Intelligence." Mind 59, no. 236 (1950): 433-460. [Ch. 1]

## U

U.S. Bureau of Labor Statistics. Occupational Outlook Handbook: "Bank Tellers," "Electricians," and "Heating, Air Conditioning, and Refrigeration Mechanics and Installers." 2024-2034 employment projections. [Chs. 14, 15]

U.S. Congress. TAKE IT DOWN Act. Pub. L. No. 119-12, signed May 19, 2025. [Ch. 9]

U.S. Congress. DEFIANCE Act. U.S. Senate passage by unanimous consent, January 2026. [Ch. 9]

U.S. Copyright Office. "Copyright and Artificial Intelligence, Part 2: Copyrightability." January 2025. [Ch. 10]

U.S. Copyright Office. "Copyright and Artificial Intelligence, Part 3: Generative AI Training." Pre-publication draft, May 2025. [Ch. 10]

U.S. Department of Agriculture, Economic Research Service. Historical farm-employment share data. [Ch. 15]

U.S. Department of Defense. Directive 3000.09, "Autonomous Weapon Systems." Updated January 25, 2023. [Ch. 11]

U.S. Department of Defense. Project Maven / Maven Smart System program materials and FY2026 budget request documentation. [Ch. 11]

U.S. Department of Health and Human Services, Office of Inspector General. "Medicare Advantage Organizations Overturned Nearly All Appealed Prior

Authorization Denials for Skilled Nursing Facility Admission, Raising Concerns About Initial Denials." June 11, 2026. VERIFIED. [Ch. 7]

U.S. Environmental Protection Agency. WaterSense program data on residential lawn and landscape irrigation. [Ch. 16]

U.S. Food and Drug Administration. "Pulse Oximeter Accuracy and Limitations: FDA Safety Communication." February 19, 2021, with subsequent draft guidance, January 2025. [Ch. 7]

U.S. Food and Drug Administration. "Clinical Decision Support Software: Guidance for Industry and Food and Drug Administration Staff." Finalized September 28, 2022; revised January 2026. [Ch. 15]

U.S. Government Accountability Office. "Facial Recognition Technology: Federal Law Enforcement Agencies Should Take Actions to Implement Training and Policies for Civil Liberties." GAO-21-518, June 2021; follow-up report, September 2023; status update, March 2024. [Ch. 12]

U.S. Senate. Vote to strip the AI regulation moratorium from H.R. 1 (One Big Beautiful Bill Act). July 1, 2025 (99-1). [Chs. 13, 17]

U.S. Senate, Permanent Subcommittee on Investigations. "Refusal of Recovery: How Medicare Advantage Insurers Have Denied Patients Access to Post-Acute Care." October 2024. [Ch. 7]

UC Berkeley School of Law, Center for Law, Energy and the Environment. Report on data center water-use disclosure gaps in California. February 2026. [Ch. 16]

UMG Recordings, Inc. et al. v. Suno, Inc. and Uncharted Labs, Inc. (Udio). U.S. District Court, Southern District of New York and District of Massachusetts, filed June 2024; Universal-Udio and Warner-Udio/Suno licensing settlements, October-November 2025. [Ch. 10]

UNESCO. "Recommendation on the Ethics of Artificial Intelligence." Adopted by the General Conference, 41st session, November 2021. [Ch. 11]

UNESCO. Miao, Fengchun, Kelly Shiohira, and Natalie Lao. "AI Competency Framework for Students." August 2024. [Ch. 18]

United Nations University Institute for Water, Environment and Health. Report projecting global AI-related data center water use by 2030. June 2026. [Ch. 16]

University of Illinois Chicago. Survey of approximately 1,500 Amazon warehouse workers on pace and safety. [Ch. 12]

Utah Legislature. Artificial Intelligence Policy Act, effective May 2024. [Ch. 13]

## V

Valley News Live / KVLY. Reporting on the Fargo Police Department press conference (Chief Dave Zibolski). March 24, 2026. [Ch. 12]

Vaswani, Ashish, Noam Shazeer, Niki Parmar, Jakob Uszkoreit, Llion Jones, Aidan N. Gomez, Lukasz Kaiser, and Illia Polosukhin. "Attention Is All You Need." arXiv:1706.03762, June 12, 2017. [Ch. 1]

## W

Washington State Department of Labor and Industries v. Amazon.com Services LLC. Citation appeal and 2024 ruling. [Ch. 12]

Weizenbaum, Joseph. "ELIZA: A Computer Program for the Study of Natural Language Communication Between Man and Machine." Communications of the ACM 9, no. 1 (1966): 36-45. [Ch. 1]

White House (The). "Winning the AI Race: America's AI Action Plan." July 2025. [Ch. 13]

White House (The). "National Policy Framework for Artificial Intelligence." March 2026. [Ch. 13]

Wouters, Olivier J., Martin McKee, and Jeroen Luyten. "Estimated Research and Development Investment Needed to Bring a New Medicine to Market, 2009-2018." JAMA 323, no. 9 (2020): 844-853. [Ch. 6]

World Economic Forum. "Future of Jobs Report 2025." January 2025. [Ch. 15]

## Z

Zee, Benny, et al. Machine-learning retinal-imaging study for autism spectrum disorder screening. eClinicalMedicine, November 2020. Zee and a co-author hold a patent on the underlying method. CONFLICT OF INTEREST DISCLOSED IN TEXT. [Ch. 6]

www.ingramcontent.com/pod-product-compliance
Lightning Source LLC
LaVergne TN
LVHW100305210926
848480LV00003B/1023
*9798990983823*